TECHNIQUES FOR VIRTUAL PALAEONTOLOGY

New Analytical Methods in Earth and Environmental Science

A new e-book series from Wiley-Blackwell

Because of the plethora of analytical techniques now available, and the acceleration of technological advance, many earth scientists find it difficult to know where to turn for reliable information on the latest tools at their disposal, and may lack the expertise to assess the relative strengths or potential limitations of a particular technique. This new series addresses these difficulties, and by virtue of its comprehensive and up-to-date coverage, provides a trusted resource for researchers, advanced students and applied earth scientists wishing to familiarise themselves with emerging techniques in their field.

Authors will be encouraged to reach out beyond their immediate speciality to the wider earth science community, and to regularly update their contributions in the light of new developments.

Written by leading international figures, the volumes in the series will typically be 75–200 pages (30,000 to 60,000 words) in length – longer than a typical review article, but shorter than a normal book. Volumes in the series will deal with:

- the elucidation and evaluation of new analytical, numerical modelling, imaging or measurement tools/techniques that are expected to have, or are already having, a major impact on the subject;
- new applications of established techniques;
- interdisciplinary applications using novel combinations of techniques.

All titles in this series are available in a variety of full-colour, searchable e-book formats. Titles are also available in an enhanced e-book edition which may include additional features such as DOI linking, high resolution graphics and video.

Series Editors

Kurt Konhauser, University of Alberta (biogeosciences)
Simon Turner, Macquarie University (magmatic geochemistry)
Arjun Heimsath, Arizona State University (earth-surface processes)
Peter Ryan, Middlebury College (environmental/low T geochemistry)
Mark Everett, Texas A&M (applied geophysics)

TECHNIQUES FOR VIRTUAL PALAEONTOLOGY

MARK D. SUTTON
Department of Earth Science and Engineering, Imperial College London, London, UK

IMRAN A. RAHMAN
School of Earth Sciences, University of Bristol, Wills Memorial Building, Bristol, UK

RUSSELL J. GARWOOD
School of Materials/School of Earth, Atmospheric and Environmental Sciences, University of Manchester, Manchester, UK

WILEY Blackwell

Registered Office
John Wiley & Sons, Ltd, The Atrium, Southern Gate, Chichester, West Sussex, PO19 8SQ, UK

Editorial Offices
9600 Garsington Road, Oxford, OX4 2DQ, UK
The Atrium, Southern Gate, Chichester, West Sussex, PO19 8SQ, UK
111 River Street, Hoboken, NJ 07030–5774, USA

For details of our global editorial offices, for customer services and for information about how to apply for permission to reuse the copyright material in this book please see our website at www.wiley.com/wiley-blackwell.

Library of Congress Cataloging-in-Publication Data

Sutton, M. D. (Mark D.), author.
  Techniques for virtual palaeontology / Mark D. Sutton, Imran A. Rahman, Russell J. Garwood.
     pages   cm
  Includes bibliographical references and index.
  ISBN 978-1-118-59113-0 (cloth)
1. Paleontological modeling.   2. Virtual reality in paleontology.   3. Paleontology–Data processing.   I. Rahman, Imran A., author.   II. Garwood, Russell J., author.   III. Title.
  QE721.2.M63S88 2014
  560.285–dc23
                                                              2013024697

A catalogue record for this book is available from the British Library.

Wiley also publishes its books in a variety of electronic formats. Some content that appears in print may not be available in electronic books.

Cover image: "Ventral view of the horseshoe crab *Dibasterium durgae* Briggs et al. 2012, from the Silurian-aged Herefordshire Lagerstätte, England. Model reconstructed through physical-optical tomography, manual registration, virtual preparation, isosurfacing and ray-tracing, using software packages SPIERS and Blender." Briggs, D.E.G., Siveter, Derek J., Siveter, David J., Sutton, M.D., Garwood, R.J, & Legg, D. 2012. Silurian horseshoe crab illuminates the evolution of arthropod limbs. P.N.A.S. 109, 15702–15705.

Set in 10/12.5pt Minion by SPi Publisher Services, Pondicherry, India

1   2014

Contents

Acknowledgements

We thank Ian Francis (Wiley-Blackwell) for initially approaching MDS to solicit this work, and Delia Sandford and Kelvin Matthews (Wiley-Blackwell) for their assistance in technical matters during its production. The following are thanked for providing suggestions to improve our first drafts: Alex Ball, Karl Bates, Robert Bradley, Jen Bright, Martin Dawson, Kate Dobson, Phil Donoghue, Peter Falkingham, Stephan Lautenschlager, Heinrich Mallison, Maria McNamara, Laura Porro, Paul Shearing and Alex Ziegler. Alan Spencer assisted with photography. We also thank the following for permission to re-use figures or for providing previously unpublished images: Karl Bates, Jason Dunlop, Cornelius Faber, Peter Falkingham, Nicolas Goudemand, Joachim Haug, Jason Hilton, Thomas Kleinteich, Heinrich Mallison, Andrew McNeil, Daniel Mietchen, Susanne Mueller, David Penney, Robert Scott, Leyla Seyfullah, David Wacey, Mark Wilson, Philip J. Withers, Florian Witzmann and Alex Ziegler. IR was funded by a NERC Postdoctoral Research Fellowship (NE/H015817/1). RG was funded by an 1851 Royal Commission Research Fellowship. Finally, we wish to thank our families, partners, friends and institutions for their forbearance with us over the long, cold winter of 2012/13, during which this book has taken shape.

1

Introduction and History

Abstract: We define virtual palaeontology as the study of three-dimensional fossils through digital visualizations. This approach can be the only practical means of studying certain fossils, and also brings benefits of convenience, ease of dissemination, and amenability to dissection and mark-up. Associated techniques fundamentally divide into surface-based and tomographic; the latter is a more diverse category, sub-divided primarily into destructive and non-destructive approaches. The history of the techniques is outlined. A long history of physical-optical studies throughout the 20th century predates the true origin of virtual palaeontology in the 1980s. Subsequent development was driven primarily by advances in X-ray computed tomography and computational resources, but has also been supplemented by a range of other technologies.

1.1 Introduction

Virtual palaeontology is the study of fossils through interactive digital visualizations, or **virtual fossils**. This approach involves the use of cutting-edge imaging and computer technologies in order to gain new insights into fossils, thereby enhancing our understanding of the history of life. While virtual palaeontological techniques do exist for handling two-dimensional data (e.g. the virtual lighting approach of Hammer et al. 2002), for most palaeontologists the field is synonymous with the study of three-dimensionally preserved material, and the term is used in this context throughout this book. Note also that the manual construction of idealized virtual models of taxa (e.g. Haug et al. 2012, Fig. 11), while very much a worthwhile undertaking, is not included in the concept of virtual palaeontology followed herein.

The majority of fossils are three-dimensional objects. While compression of fossils onto a genuinely two-dimensional plane does of course occur (Figure 1.1a), it is the exception, and in most preservational scenarios at

Techniques for Virtual Palaeontology, First Edition. Mark D. Sutton,
Imran A. Rahman and Russell J. Garwood.
© 2014 John Wiley & Sons, Ltd. Published 2014 by John Wiley & Sons, Ltd.

Figure 1.1 Dimensionality in fossils: (a) Completely two-dimensional graptolite fossils; genuinely two-dimensional fossils such as this are the exception. (b) A three-dimensionally preserved trilobite cephalon; most fossils exhibit at least partial three-dimensional preservation. Scale bars are 10 mm. Both specimens are from Lower Ordovician, Wales.

least an element of the original three-dimensionality is retained (Figure 1.1b). Three-dimensional preservation retains more morphological information than true two-dimensional modes, but typically this information is problematic to extract. Isolation methods, of which several exist, are one solution. Fossils may simply 'drop out' or be naturally washed out of rocks; wet-sieving of poorly consolidated sediments mimics this process. Specimens may also be extracted chemically, for example, by dissolving the matrix (e.g. Aldridge 1990). These approaches are effective where applicable, but are prone to losing associations between disarticulated or weakly connected parts of fossils, and to damaging delicate structures. Specimens can also be physically 'prepared' out using needles, drills or gas-jet powder abrasive tools (e.g. Whybrow and Lindsay 1990); while usually preserving associations, this approach may also damage delicate structures, scales poorly to small specimens, and cannot always expose all of a specimen. Finally, isolation of a fossil only provides access to its surface.

Correctly chosen, virtual palaeontological techniques can overcome many of the disadvantages of physical isolation methods, and bring many novel advantages too. Virtual specimens are typically more convenient to work with, requiring only a computer rather than expensive and lab-bound microscopes. They allow for virtual dissection and sectioning, where parts of the specimen can be isolated for clarity without fear of damage. They allow for mark-up, typically in the form of colour applied to discrete anatomical elements, which can greatly increase the ease of interpretation. They can be used as the basis for quantitative studies of functional morphology, such as **finite-element analysis** of stress and strain (e.g. Rayfield 2007), or hydrodynamic flow modelling (e.g. Shiino et al. 2009). Finally, as virtual specimens are simply computer files, they can be easily copied and disseminated to interested parties, facilitating collaborative analysis and publication.

Despite all these advantages, virtual palaeontology is not as widely used as it might be; one possible reason is that the techniques involved are perceived as 'difficult', and while there is no lack of technical detail available on individual techniques, no in-depth treatment and comparison of all available

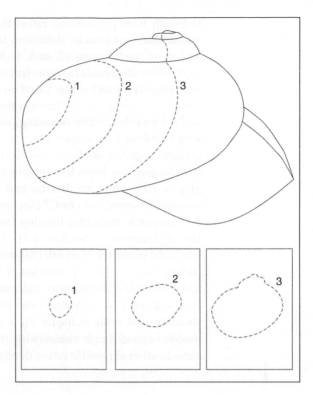

Figure 1.2 Tomography. Three parallel and evenly spaced serial tomograms (1–3) through an idealized gastropod fossil, and the resultant tomographic dataset. Modified from Sutton (2008, Fig. 1). Reproduced with permission of The Royal Society of London.

techniques exists, which can make the field intimidating to those entering it for the first time. This book aims to overcome this issue. It is intended to provide those interested in doing palaeontology through virtual methods, or in interpreting virtual data provided by other workers, with background theoretical knowledge and practical grounding. In particular, it aims to provide palaeontologists with the information they need to select an appropriate methodology for any particular study, to understand the pitfalls and limitations of each technique, and to provide suggestions for carrying out work with maximal efficiency. Theoretical concepts are covered with the intention of providing scientists with sufficient depth of understanding to develop and modify techniques, where appropriate.

Virtual palaeontological data-capture techniques can be divided most fundamentally into (a) tomographic (slice-based) approaches, and (b) surface-based approaches. **Tomography** is the study of three-dimensional structures through a series of two-dimensional parallel 'slices' through a specimen (Figure 1.2). In tomography, an individual slice-image is termed a **tomogram**, and a complete set of tomograms is (herein) termed a **tomographic dataset**. Any device capable of producing tomograms is a **tomograph**. Note that while the definition of tomography given above is the original one (derivation is from the Greek *tomos* – section, cut, slice and *graphein* – writing, imaging, study), in recent years this term has often been restricted to techniques where virtual tomograms are computed indirectly from **projections**, rather than imaged in a direct way. However, we consider our broader definition to be both more historically accurate and more useful, with all such techniques sharing much in common, especially with regards to reconstruction methodology. The term we prefer for tomographic techniques based on computation of virtual tomograms is **computed tomography**. Tomography can be divided into (a) destructive and (b) non-destructive (scanning) methodologies. The former include the long-established techniques of **serial grinding**, sawing, **slicing**, etc. (here grouped together as **physical-optical tomography**, Section 2.2), together with focused ion-beam tomography (Section 2.3). Non-destructive tomographic techniques are diverse, and include the many variants of **X-ray computed tomography** or CT (Section 3.2), **neutron tomography** (Section 3.3), magnetic resonance imaging (Section 3.4), and **optical tomography** (**serial focusing** – Section 3.5). Surface-based techniques are those where the geometry of an external surface is digitized in some fashion; they include **laser scanning** (Section 4.2), **photogrammetry** (Section 4.3) and **mechanical digitization** (Section 4.4). This book concludes with an examination of the techniques and software available for specimen reconstruction and study (Chapter 5), a review of the applications of virtual models beyond simple visualization (Chapter 6), and a final overview and consideration of possible future developments (Chapter 7).

1.2 Historical Development

Virtual Palaeontology, in the sense used in this book, began in the early 1980s when the emerging medical technology of X-ray computed tomography was first applied to vertebrate fossils. The power of tomography to document and reconstruct three-dimensionally preserved material has, however, long been recognized, and modern techniques have a lengthy prehistory of physical-optical tomography (*sensu* Section 2.2), combined in some cases with non-computerized visualization techniques.

1.2.1 Physical-Optical Tomography in the 20th Century

Palaeontological tomography was introduced in the first years of the 20th century by the eccentric Oxford polymath William J. Sollas, who noted the utility of serial sectioning in biology and realized that serial grinding could provide similar datasets from palaeontological material. His method (Sollas 1903) utilized a custom-made serial-grinding tomograph capable of operating at 25 μm intervals, photography of exposed surfaces, and manual tracing from glass photographic plates. Sollas applied this approach with considerable zeal to a wide range of fossil material, and was able to demonstrate the fundamental utility and resolving power of tomography to a broad audience. He also described (Sollas 1903) a physical-model visualization technique in which tomograms were traced onto thin layers of beeswax which could then be cut to reproduce the original slice, stacked together and weakly heated to fuse them into a cohesive model. A quick-and-dirty approach to model-making, using glued cardboard slices rather than fused wax, was also in early use; while documentation is lacking, this appears also to be traceable back to Sollas.

Sollas was primarily a vertebrate palaeontologist, and it was in this field that his methods first became widely accepted, most notably in the seminal studies of Stensiö (1927) on the cranial anatomy of Devonian fish. From the mid-20th century, however, serial grinding became a well-established palaeontological technique, and was applied to a very wide range of fossil vertebrates, invertebrates, and plants. These applications are far too numerous to cite, but an excellent example of a group whose students embraced it with some degree of fervour is the Brachiopoda. Brachiopods are often preserved three-dimensionally and articulated with valves firmly closed, concealing taxonomically and palaeobiologically informative internal structures such as lophophore supports; following the pioneering work of Muir-Wood (1934), the use of manually traced serial sections to document these structures has become almost ubiquitous.

A range of serial-grinding tomographs, for the most part custom-built devices, have been used since Sollas's work (e.g. Simpson 1933; Croft 1950; Ager 1965; Sutton et al. 2001b); these have varied substantially in complexity, degree of automation, maximum specimen size and minimum grind-interval, although none have substantially improved on the original machine in the latter respect. Two major variants on the technique have also been important, both responses to the destructive nature of serial grinding. Firstly, **acetate peels** (Walton 1928, see Galtier and Phillips 1999 for a more modern treatment) have been widely adopted as a means of data capture, especially but not exclusively in palaeobotany. **Peels** provide a permanent record of mineralogy and can be combined with staining techniques to increase contrast between certain types of material; they have thus been viewed as superior to mere photography of surfaces. Peels do, however, bring a peculiar set of problems of their own (see Section 2.2.2.3), and their use has unfortunately rendered many historical datasets ill-suited to modern visualization methods. Secondly, **serial sawing** using fine annular or diamond-wire saws (Kermack 1970) became popular for larger fossils such as vertebrates in the latter quarter of the 20th century, as it allowed retention of original material (albeit at the cost of an increase in minimum tomogram spacing).

While physical-optical tomography was commonplace in the 20th century, physical model-making noticeably fell out of favour, considered perhaps to be too laborious and of doubtful scientific utility. Students of particular groups (e.g. brachiopods) became sufficiently familiar with tomograms to be able to integrate them into mentally conceived three-dimensional representations, and the potential benefits of being able to directly communicate these visualizations beyond the *cognoscenti* were arguably overlooked. Reconstructions from tomographic data, where published, typically took the form of idealized pictorial or diagrammatic representations from such mentally assembled models; while aesthetically pleasing and often gratifyingly simplified (for an example from palaeobotany see the cupule reconstructions of Long 1960), this form of reconstruction lacked objectivity. That said, physical models were undoubtedly difficult to assemble, fragile, difficult to transport and hard to work with; while some workers continued to use them (e.g. Jefferies and Lewis 1978), truly effective visualization was not eventually achieved until the advent of interactive virtual fossils at the start of the 21st century.

1.2.2 The CT Revolution

Tomography in palaeontology has seen an enormous rise in uptake in recent years – Figure 1.3 provides a graphical representation of the use of the term 'tomography' in the palaeontological literature. It shows a fairly steady rise for the 30 years between 1975 and 2005 (the drop in 1996 is probably a methodological artefact of the way the literature was indexed), followed by an upswing that is, to say the least, eye-catching. This phenomenon is, for the most part, a result of the increasing availability and popularity of X-ray CT, and we refer to it herein as the **CT revolution**. X-ray computed axial tomography (CT or CAT scanning) is a technology that arose as an advanced form of medical radiography in the early 1970s, taking advantage of the increasing availability of computing power together with technical and algorithmic advances. CT, its history and its derivatives are described in more detail in

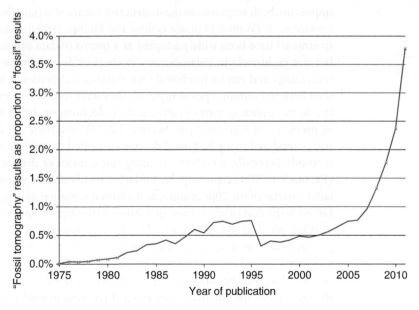

Figure 1.3 Relative increase in the importance of tomography in palaeontology from 1975 to 2011, as calculated by the ratio of publications including 'fossil' and 'tomography' to those only including 'fossil'. Data from Google Scholar (scholar.google.co.uk), July 2012. While these data inevitably include biases, they give a clear indication of trends. Note that the step down in 1996 is best interpreted as an artefact of the search engine.

Section 3.2. Many types of fossil material have long been known to be amenable to X-ray analysis (i.e. to have high contrast between fossil and matrix in terms of X-ray attenuation), and this form of **non-destructive tomography** thus clearly had palaeontological potential. Early machines were limited in availability and resolution, however, so it was not until 1982 that CT was first applied to vertebrate fossil material (Tate and Cann 1982, see also Conroy and Vannier 1984). Medical development of CT was accompanied by parallel development of visualization tools, and thus by the time these early studies were undertaken three-dimensional digital models, albeit in a somewhat limited form, could be reconstructed from the data. Arguably, the first high-profile palaeontological use of the technology was in a restudy of *Archaeopteryx* (Haubitz et al. 1988), and since the 1990s the technology has become increasingly commonplace for the study of the relatively large specimens typical of vertebrate palaeontology, many of which are suited for the range of scales handled by the readily available medical scanners. Serious study of invertebrate and other smaller fossils using CT did not begin until the 21st century (although see Hamada et al. 1991), with the advent of **X-ray microtomography** (XMT). Developed initially by Elliot and Dover (1982), XMT systems work on smaller scales, typically with resolutions down to a few microns. The palaeontological pioneers of XMT worked in the University of Texas High-Resolution X-ray Computed Tomography Facility (see e.g. Rowe et al. 2001 and www.digimorph.org), but the increasing availability of relatively low-cost laboratory or even desktop-scale scanners in recent years has resulted in a profusion of studies using XMT. Finally, the advent of X-ray tomography beamlines at third-generation synchrotrons (see e.g. Donoghue et al. 2006; Tafforeau et al. 2006) has provided facilities for extremely high-resolution and high-fidelity tomographic study of palaeontological material. Particularly in combination with methodological advances such as phase-contrast imaging, these facilities have enabled the study of otherwise intractable material in an unparalleled level of detail.

1.2.3 Modern Physical-Optical Tomography

Although the CT revolution has hugely increased usage of tomographic methods, it has not entirely swept away traditional physical-optical methods; rather, these have enjoyed a limited resurgence. Despite their destructive nature they remain, for some material, the most cost-efficient or even the only practical means of data recovery. The study of the invertebrate fossils of the Silurian Herefordshire Lagerstätte (Briggs et al. 2008) has provided the best example of this resurgence, demonstrating in a series of publications the power of serial-grinding tomography married to modern digital photography; Watters and Grotzinger (2001) provide a contemporaneous example of similar techniques applied to different material. The nature of existing physical-optical datasets, typically relatively sparse in terms of tomogram spacing, drove early experimentation with vector-based digital visualization (e.g. Chapman 1989; Herbert 1999), where manually or automatically traced structures were surfaced to produce reconstructions which were crude but low in polygon count and hence easily rendered on available hardware. Other ingenious but somewhat idiosyncratic approaches to visualization were also tried (e.g. Hammer 1999), but it was only with the application of the more medically mainstream approach of **isosurface** generation and rendering (see Chapter 5) to Herefordshire data by Sutton et al. (2001a, b) that genuinely high-fidelity virtual models from physical-optical data began to appear, the key ingredient simply being the collection of a large number of closely spaced tomograms. The isosurface approach has been the primary visualization tool used for all palaeontological tomographic datasets since that study, although **direct volume rendering** (e.g. Hagardorn et al. 2006) and **vector surfacing** (e.g. Kamenz et al. 2008) have found occasional applications.

1.2.4 Other Modern Tomographic Techniques

Other approaches to palaeontological tomography exist, of course, and are detailed in this book (see Section 1.1); they include **magnetic resonance imaging (MRI)**, neutron tomography, optical tomography, and **focused ion beam (FIB) tomography**. All could fairly be described as niche techniques, and their history of application is, in each case, short. MRI is a medical scanning technology that was initially developed during the 1970s; while MRI tomograms are typically lower resolution than those generated by CT, radiation doses are lower, and for medical samples data acquisition can be faster and tissue differentiation better. None of these advantages are especially relevant to palaeontological material, however, and MRI often performs poorly on solid materials. Applications have hence been rare and primarily experimental in nature (Mietchen et al. 2008, although see Gingras et al. 2002; Clark et al. 2004 for practical applications). Neutron tomography utilizes neutron beams to perform tomography in a manner analogous to CT. Some studies have demonstrated limited utility, particularly in fossils preserving organic compounds (Schwarz et al. 2005; Winkler 2006), and the relatively weak absorption of neutrons by metal-rich rocks theoretically allows large and dense specimens, opaque to X-ray beams, to be studied. However, the relatively low resolution of the technique together with the limited number of facilities at which it can be undertaken have militated against a broad uptake. Optical tomography or serial focusing, typically but not exclusively using **confocal microscopy**, provides a very high-resolution non-destructive approach to tomographic data capture, albeit only for translucent samples and only on small scales. The optical techniques concerned have a long history, confocal microscopy originating in the late 1980s and less precise serial-focusing methods having existed long before; however, while confocal microscopy was first applied to fossils in the 1990s (e.g. Scott and Hemsley 1991; O'Connor 1996), applications of any optical tomography techniques to palaeontological material since have been sporadic (e.g. Ascaso et al. 2003; Schopf et al. 2006, Kamenz et al. 2008). Finally, focused ion beam (FIB) microscopes were developed primarily for use in material science in the late 1970s (see e.g. Phaneuf 1999); while they were originally used for imaging, the ion beam can also mill material, and hence they can be employed, somewhat laboriously, to perform nano-scale tomography. Although a smattering of studies has been published in recent years (e.g. Schiffbauer and Xiao 2009; Wacey et al. 2012), this approach has yet to see widespread application to fossil material.

1.2.5 Surface-Based Techniques

Surface-based digitization techniques represent an entirely different approach to virtual palaeontology (see Section 1.1); rather than relying on tomograms, these approaches digitize the topography of the surface of a specimen, and can also capture surface colour. While obviously inappropriate for looking inside physical objects, they represent a powerful set of techniques for performing virtual palaeontology on fossils where the surface morphology represents all or most of the preserved information. While a substantial portion of this book is devoted to these methods, their history of usage in palaeontology is brief.

Contact or mechanical digitization involves the use of a robotic arm equipped with sensors that can record the position of a tip in three-dimensional space; an operator can use this device to collect surface points over an object. Developed in the 1990s for a variety of digitization applications, this approach has been sporadically applied in palaeontology in the 21st century (Wilhite 2003; Mallison et al. 2009), although only to vertebrate fossils.

The majority of surface-based digitization has instead made use of laser scanning, a set of techniques where the reflection of a scanned laser-beam from a surface is used to record surface topography at distance. The technology was first commercialized in the 1980s for capturing human faces and later entire bodies for the animation industry, and the first relatively portable devices capable of rapid and precise scanning became available in the late 1990s; since then they have become increasingly cheaper and better specified. The first palaeontological application was by Lyons et al. (2000) in a study of part of a dinosaur skull; subsequently, a flurry of studies have used this approach on a range of fossils including vertebrates (Bates et al. 2009), footprints (Bates et al. 2008) and Ediacaran problematica (see e.g. Antcliffe and Brasier 2011). The technique is also in curatorial use for major museum-based digitization initiatives such as the *GB/3D type fossils online* project (Howe 2012), which, at the time of writing, is undertaking laser-scan digitization of a substantial proportion of all UK-held-type fossil specimens.

The other important surface-based approach to digitization is photogrammetry, in which three-dimensional models are assembled from a series of two-dimensional photographs of an object. Digital photogrammetry has a long pre-history that can be traced back to the origins of photography, and analogue photogrammetry has long been important, in cartography in particular (see e.g. Kraus 2007). The widespread use of stereo-pair images in palaeontology to provide a form of three-dimensional model can also be seen as a forerunner of true photogrammetry-based virtual palaeontology. As techniques have matured and digital photogrammetry has become available, in which models are automatically constructed direct from digitally captured images, a rapid expansion of applications has taken place; photogrammetry is now widely used in forensics and archaeology, for example. Palaeontological applications have hitherto been few, and predominantly concerned with dinosaur tracks (e.g. Breithaupt and Matthews 2001; Bates et al. 2009). However, recent developments in photogrammetric software (see Falkingham 2012) suggest that photogrammetry can be at least as effective as laser scanning in some palaeontological contexts, and the method can be expected to become increasingly important in the near future.

1.2.6   Historical Summary

The history of virtual palaeontology is relatively short when considered in its narrowest form. However, when considered with its precursors and related methods, it shows a long-standing appreciation in the palaeontological community of the value of three-dimensional data and models, despite the difficulties in actually obtaining them using older methods. The last decade has seen a remarkable rise both in the number of studies using virtual palaeontological techniques and in the breadth of techniques employed; this outpouring represents not simply the exploitation of newly available opportunities, but also the satisfaction of a long-present hunger amongst palaeontologists. Virtual palaeontology enables us to work with three-dimensional fossils not so much in a 'way that we never knew we could', more a 'way that we always thought we should, but didn't know how to'.

References

Ager, D.V. (1965) Serial grinding techniques. In: Kummel, B. & Raup, D. (eds), *Handbook of Palaeontological Techniques*, pp. 212–224. H. Freeman, San Fransisco.

Ascaso, C., Wierzchos, J., Corral, J.C., et al. (2003) New applications of light and electron microscopic techniques for the study of microbiological inclusions in amber. *Journal of Paleontology*, **77** (6), 1182–1192.

Aldridge, R.J. (1990) Extraction of microfossils. In: Briggs, D.E.G. & Crowther, P.R. (eds), *Palaeobiology: A Synthesis*, pp. 502–504. Blackwell, Oxford.

Antcliffe, J.B. & Brasier, M.D. (2011) Fossils with little relief: using lasers to conserve, image, and analyse the Ediacara biota. In: Laflamme, M., Schiffbauer, J.D. & Dornbos, S.Q. (eds), *Quantifying the Evolution of Early Life: Numerical Approaches to the Evaluation of Fossils and Ancient Ecosystems*, pp. 223–240. Springer, Dordrecht.

Bates K.T., Manning, P.L., Hodgetts D., et al. (2009) Estimating mass properties of dinosaursusing laser imaging and 3D computer modelling. *PLoS ONE*, **4 (2)**, e4532.

Bates K.T., Rarity F., Manning P.L., et al. (2008) High-resolution LiDAR and photogrammetric survey of the Fumanya dinosaur tracksites (Catalonia): implications for the conservation and interpretation of geological heritage sites. *Journal of the Geological Society of London*, **165 (1)**, 115–127.

Breithaupt, B.H. & Matthews, N.A. (2001) Preserving paleontological resources using photogrammetry and geographic information systems. In: Harmon, D. (ed) *Crossing Boundaries in Park Management: Proceedings of the 11th Conference on Research and Resource Management in Parks and Public Lands*, pp. 62–70. The George Wright Society, Hancock.

Briggs, D.E.G., Siveter, David J., Siveter, Derek J., et al. (2008) Virtual fossils from a 425 million-year-old volcanic ash. *American Scientist*, **96 (6)**, 474–481.

Chapman, R.E. (1989) Computer assembly of serial sections. In: Feldmann, M., Chapman, R. & Hannibal, J.T. (eds), *Paleotechniques*, pp. 157–164. Special Publication 4, Paleontological Society, Boulder.

Clark, N.D.L., Adams, C., Lawton, T., et al. (2004) The Elgin marvel: using magnetic resonance imaging to look at a mouldic fossil from the Permian of Elgin, Scotland, UK. *Magnetic Resonance Imaging*, **22 (2)**, 269–273.

Croft, W.N. (1950) A parallel grinding instrument for the investigation of fossils by serial sections. *Journal of Paleontology*, **24 (6)**, 693–698.

Conroy, G.C. & Vannier, M.W. (1984) Noninvasive three-dimensional computer imaging of matrix-filled fossil skulls by high-resolution computed tomography. *Science*, **226 (4673)**, 456–458.

Donoghue, P.C.J., Bengtson, S., Dong, X. et al. (2006) Synchrotron X-ray tomographic microscopy of fossil embryos. *Nature*, **442 (7103)**, 680–683.

Elliot, J.C. & Dover, S.D. (1982) X-ray microtomography. *Journal of Microscopy*, **126 (2)**, 211–213.

Falkingham, P.L. (2012) Acquisition of high resolution 3D models using free, open-source, photogrammetric software. *Palaeontologia Electronica*, **15 (1)**, 1T.

Galtier, J. & Phillips, T. (1999) The acetate peel technique. In: Jones, T.P. & Rowe, N.P. (eds), *Fossil Plants and Spores*, pp. 67–70. The Geological Society, London.

Gingras, M.K., MacMillan, B., Balcom, B.J., et al. (2002) Using magnetic resonance imaging and petrographic techniques to understand the textural attributes and porosity distribution in Macaronichnus-burrowed sandstone. *Journal of Sedimentary Research*, **72 (4)**, 552–558.

Hagadorn, J.W., Xiao, S., Donoghue, P.C.J., et al. (2006) Cellular and subcellular structure of neoproterozoic animal embryos. *Science*, **314 (5797)**, 291–294.

Hamada, T., Tateno, S. & Suzuki, N. (1991) Three dimensional reconstruction of fossils with X-ray and computer graphics. *Scientific Papers of the College of Arts and Sciences. The College of Arts and Sciences (Kyoyo-Gakubo), The University of Tokyo, Tokyo*, **41**, 107–118.

Hammer, Ø. (1999) Computer-aided study of growth patterns in tabulate corals, exemplified by *Catenipora heintzi* from Ringerike, Oslo Region. *Norsk Geologisk Tidsskrift*, **79 (4)**, 219–226.

Hammer, Ø., Bengtson, S., Malzbender, T., et al. (2002) Imaging fossils using reflectance transformation and interactive manipulation of virtual light sources. *Palaeontologia Electronica*, **5 (4)**, 9A.

Haubitz, B., Prokop, M., Doehring, W., et al. (1988) Computed tomography of *Archaeopteryx*. *Palaeobiology*, **14 (2)**, 206–213.

Haug, J.T., Briggs, D.E.G. & Haug, C. (2012) Morphology and function in the Cambrian Burgess Shale megacheiran arthropod *Leanchoilia superlata* and the application of a descriptive matrix. *BMC Evolutionary Biology*, **12**, 162.

Herbert, M.J. (1999) Computer-based serial section reconstruction. In: Harper, D.A.T. (ed), *Numerical Palaeobiology: Computer-Based Modelling and Analysis of Fossils and Their Distributions*, pp. 93–126. Wiley, Chichester.

Howe, M. (2012) GB/3D Type fossils online (abstract). In: Young, J. & Dunkley Jones, T. (eds), *Big Palaeontology – Lyell Meeting 2012*, p. 19. The Geological Society, London.

Jefferies, R.P.S. & Lewis, D.N. (1978) The English Silurian fossil *Placocystites forbesianus* and the ancestry of the vertebrates. *Philosophical Transactions of the Royal Society, B*, **282 (990)**, 205–323.

Kamenz, C., Dunlop, J.A., Scholtz, G., et al. (2008) Microanatomy of early Devonian book lungs. *Biology Letters*, **4 (2)**, 212–215.

Kermack, D.M. (1970) True serial-sectioning of fossil material. *Biological Journal of the Linnean Society*, **2 (1)**, 47–53.

Kraus, K. (2007) *Photogrammetry: Geometry from Images and Laser Scans* (2nd edition). Walter de Gruyter, Berlin.

Long, A.G. (1960) *Stamnostoma huttonense* gen. et sp. nov., a pteridosperm seed and cupule from the Calciferous Sandstone series of Berwickshire. *Transactions of the Royal Society of Edinburgh*, **64**, 201–215.

Lyons, P.D., Rioux, M. & Patterson, R.T. (2000) Application of a three-dimensional color laser scanner to paleontology: an interactive model of a juvenile *Tylosaurus* sp. basisphenoid-basioccipital. *Palaeontologia Electronica*, **3 (2)**, 4A.

Mallison, H., Hohloch, A. & Pfretzschner, H. (2009) Mechanical digitizing for paleontology – new and improved techniques. *Palaeontologia Electronica* **12 (2)**, 4T.

Mietchen, D., Aberhan, M., Manz, B., et al. (2008) Three-dimensional magnetic resonance imaging of fossils across taxa. *Biogeosciences*, **5 (1)**, 25–41.

Muir-Wood, H.M. (1934) On the internal structure of some mesozoic Brachiopoda. *Philosophical Transactions of the Royal Society of London. Series B*, **223 (494–508)**, 511–567.

O'Connor, B. (1996) Confocal laser scanning microscopy: a new technique for investigating and illustrating fossil Radiolaria. *Micropalaeontology*, **42 (4)**, 395–402.

Phaneuf, M.W. (1999) Applications of focused ion beam microscopy to materials science specimens. *Micron*, **30 (3)**, 277–288.

Rayfield, E.J. (2007) Finite element analysis and understanding the biomechanics and evolution of living and fossil organisms. *Annual Review of Earth and Planetary Sciences*, **35**, 541–576.

Rowe, T.B., Colbert, M., Ketcham, R.A., et al. (2001) High-resolution X-ray computed tomography in vertebrate morphology. *Journal of Morphology*, **248 (3)**, 277–278.

Scott, A.C. & Hemsley, A.R. (1991) A comparison of new microscopical techniques for the study of fossil spore wall ultrastructure. *Review of Palaeobotany and Palynology*, **67 (1–2)**, 133–139.

Schiffbauer, J.D. & Xiao, S. (2009) Novel application of focused ion beam electron microscopy (FIB-EM) in preparation and analysis of microfossil ultrastructures: a new view of complexity in early eukaryotic organisms. *Palaios*, **24 (9)**, 616–626.

Schopf, J.W., Tripathia, A.B. & Kudryavstev, A.B. (2006) Three-dimensional confocal optical imagery of Precambrian microscopic organisms. *Astrobiology*, **6 (1)**, 1–16.

Schwarz, D., Vontobel, P.L., Eberhard, H., et al. (2005) Neutron tomography of internal structures of vertebrate remains: a comparison with X-ray computed tomography. *Paleontologia Electronica*, **8 (2)**, 30A.

Simpson, G.G. (1933) A simplified serial sectioning technique for the study of fossils. *American Museum Novitates*, **634**, 1–6.

Shiino, Y., Kuwazuru, O. & Yoshikawa, N. (2009) Computational fluid dynamics simulations on a Devonian spiriferid *Paraspirifer bownockeri* (Brachiopoda): generating mechanism of passive feeding flows. *Journal of Theoretical Biology*, **259** (**1**), 132–141.

Sollas, W.J. (1903) A method for the investigation of fossils by serial sections. *Philosophical Transactions of the Royal Society of London, B*, **196 (214–224)**, 259–265.

Stensiö, E.A. (1927) The Downtonian and Devonian vertebrates of Spitzbergen. *Skrifter om Svalbard og Nordishavet*, **12**, 391.

Sutton, M.D. (2008) Tomographic techniques for the study of exceptionally preserved fossils. *Proceedings of the Royal Society B*, **275**, 1587–1593.

Sutton, M.D., Briggs, D.E.G., Siveter, David J., et al. (2001a) An exceptionally preserved vermiform mollusc from the Silurian of England. *Nature*, **410 (6827)**, 461–463.

Sutton, M.D., Briggs, D.E.G., Siveter, David J., et al. (2001b) Methodologies for the visualization and reconstruction of three-dimensional fossils from the Silurian Herefordshire Lagerstätte. *Palaeontologia Electronica*, **4 (1)**, 1A.

Tafforeau, P., Boistel, R., Boller, E., et al. (2006) Applications of X-ray synchrotron microtomography for non-destructive 3D studies of paleontological specimens. *Applied Physics A*, **83 (2)**, 195–202.

Tate, J.R. & Cann, C.E. (1982) High-resolution computed tomography for the comparative study of fossil and extant bone. *American Journal of Physical Anthropology*, **58 (1)**, 67–73.

Wacey, D., Menon, S., Green, L., et al. (2012) Taphonomy of very ancient microfossils from the ~3400 Ma Strelley pool formation and ~1900 Ma Gunflint formation: new insights using a focused ion beam. *Precambrian Research*, **220–221**, 234–250.

Walton, J. (1928) A method of preparing sections of fossil plants contained in coal balls or in other types of petrifaction. *Nature*, **122 (3076)**, 571.

Watters, W.A. & Grotzinger, J.P. (2001) Digital reconstruction of calcified early metazoans, terminal Proterozoic Nama Group, Namibia. *Paleobiology*, **27 (1)**, 159–171.

Whybrow, P.J. & Lindsay, W. (1990) Preparation of macrofossils. In: Briggs, D.E.G. & Crowther, P.R. (eds) *Palaeobiology: A Synthesis*, pp. 500–502. Blackwell, Oxford.

Wilhite, R. (2003) Digitizing large fossil skeletal elements for three-dimensional applications. *Palaeontologia Electronica*, **5 (2)**, 4A.

Winkler, B. (2006) Applications of neutron radiography and neutron tomography. *Reviews in Mineralogy and Geochemistry*, **63 (1)**, 459–471.

Further Reading/Resources

Garwood, R.J., Rahman, I.R. & Sutton, M.D. (2010) From clergymen to computers – the advent of virtual palaeontology. *Geology Today* **26**(3), 96–100.

Kak, A.C. & Slaney, M. (2001) *Principles of Computerized Tomographic Imaging*. Society of Industrial and Applied Mathematics, Philadelphia.

Sutton, M.D. (2008) Tomographic techniques for the study of exceptionally preserved fossils. *Proceedings of the Royal Society B*, **275**, 1587–1593.

Zollikofer, C.P. & Ponce de Leon, M. (2005) *Virtual Reconstruction: A Primer in Computer-Assisted Paleontology and Biomedicine*. Wiley, Chichester.

2 Destructive Tomography

Abstract: Destructive tomography involves physical exposure of surfaces; all methods are time-consuming and damage or destroy specimens, but can produce high-fidelity reconstructions. Physical-optical methods involve surface exposure (through grinding, sawing or slicing), and optical imaging (photography or tracing, sometimes via acetate peels). Grinding usually produces the best results, but is maximally destructive. Slicing can produce a high tomogram-frequency, but is difficult and may distort slices. Sawing involves a low tomogram-frequency; it is only appropriate for large specimens. Direct photography is normally the preferred imaging option. Imaging via acetate peels is not recommended. Other considerations for physical-optical tomography are discussed, and case studies are provided. FIB tomography uses a focused ion beam microscope to precisely mill material from a sample, exposing surfaces that can be imaged as tomograms using electron imaging. Very high tomogram-frequencies can be obtained, and compositional data can be recorded. FIB tomography is, however, limited to very small specimens.

2.1 Introduction

Destructive tomography includes all forms of tomography in which the specimen is at least partially destroyed in the course of the production of the tomographic dataset. In these techniques, tomograms are exposed physically (either by grinding, sawing, slicing or ion-beam milling), and are then imaged in some way (e.g. by photography).

Approaches vary greatly in detail and relative merits, as discussed later, but share certain properties that can sensibly be discussed *in toto*. Most obvious of these is their undesirable destructive nature. While destructive tomography can be defended as the conversion of a specimen from physical into digital form, rather than simply its destruction, this conversion process

Techniques for Virtual Palaeontology, First Edition. Mark D. Sutton,
Imran A. Rahman and Russell J. Garwood.
© 2014 John Wiley & Sons, Ltd. Published 2014 by John Wiley & Sons, Ltd.

is necessarily imperfect. Practical limitations on imaging resolution and inter-tomogram spacing will result in data loss, and imaging of any form cannot capture all information contained in an exposed surface – optical imaging, for instance, will not fully capture all mineralogical data. Attempts to generate a more physical record of an exposed surface in the form of acetate peels bring their own difficulties (see Section 2.2.2.3). Errors and mishaps in the preparation of a tomographic dataset using one of these techniques are also inevitable; while care can be and should be taken to minimize them, it is difficult to entirely eliminate data loss through mischance. Destructive tomography, even where carried out with the utmost care and using the best available tomographs and imaging techniques, always precludes the application of some future and potentially better data-extraction technique.

All destructive techniques are also time-consuming to employ, and most are labour-intensive. While in many cases the tomographs themselves are inexpensive, labour requirements can result in high total costs of specimen preparation. Additionally, and in contrast to most non-destructive techniques, destructively gathered tomographic datasets almost always require **registration** (alignment) prior to three-dimensional visualization (see Section 5.2.3). This step can also be time-consuming.

Despite these generic caveats, destructive tomography remains the best option for the three-dimensional study of some fossils. Certain types of material are simply not easily amenable to any non-destructive tomographic techniques – specimens might show insufficient X-ray or neutron attenuation contrast for X-ray computed tomography (CT) or neutron tomography (see Sections 3.2 and 3.3), have insufficient hydrogen content for magnetic resonance imaging (see Section 3.4), and be too opaque for optical tomography (see Section 3.5). The fossils of the Herefordshire Lagerstätte (Briggs et al. 1996) provide just such an example. Image capture from exposed surfaces also facilitates data-rich imaging modes; colour photography, for instance, can capture subtleties of composition not evident in X-ray CT, and focused ion beam (FIB) techniques (see Section 2.3) enable compositional, chemical and crystallographic mapping of surfaces. Finally, for some historical specimens destructive physical-optical tomography may have already been performed, and a palaeontologist wishing to reconstruct these specimens will have no choice other than to use the existing datasets.

The problem of specimen destruction always remains, but is of less concern where material is abundant. Nonetheless, physical-optical techniques (see Section 2.1) have even been used on singleton specimens (e.g. Sutton et al. 2012b). Note that the International Commission on Zoological Nomenclature code (ICZN 1999) does not forbid the use of destroyed specimens as holotypes for new species, article 73.1.4 stating that 'Designation of an illustration of a single specimen as a holotype is to be treated as designation of the specimen illustrated; the fact that the specimen no longer exists or cannot be traced does not of itself invalidate the designation'.

2.2   Physical-Optical Tomography

The term physical-optical tomography was introduced by Sutton (2008) to encompass a range of 'traditional' destructive tomographic techniques in which surfaces are physically exposed and imaged simply through their optical properties; it is roughly analogous to the traditional term 'serial sectioning', although slightly broader and more precise in scope. Physical-optical techniques have a long history of use in palaeontology, as discussed in Chapter 1. They are best sub-categorized by separately considering first the range of methods that can be used to physically expose the tomographic surfaces, and second those used to visually record surfaces.

2.2.1   Approaches to Surface Exposure

There are essentially three ways in which serial exposure of tomographic surfaces can be achieved; grinding, sawing or slicing. In the case of grinding, the procedure discussed later (for single-slice removal) must be alternated with an imaging step; for sawing and slicing the entire surface exposure 'run' is carried out first, and resultant surfaces then separately imaged.

2.2.1.1   Grinding

Grinding (or lapping) has been the most important of these approaches historically, and it remains the most widely employed today. Specimens are positioned against an abrasive surface and some form of motion is employed to physically grind away a small thickness of material, which might range from a few micrometres to several millimetres. Details of methodology are many and various. In its simplest form, a specimen may be manually ground by firm and repetitive circular or figure-of-eight motion against a glass plate covered with abrasive powder, although this approach offers no precise control over or measurement of the thickness of material removed. It also risks the violation of the 'parallel tomograms' requirement for visualization (see Table 5.1), as uneven pressure can consistently bias the grinding towards one side of the specimen. These limitations can be overcome by more sophisticated grinding tomographs in which specimens are clamped or mounted to maintain a consistent orientation to the grinding surface, and in which the position of the specimen can be carefully controlled with respect to the grinding surface to allow precise removal of known thicknesses of material. Many such tomographs have been employed through the long history of this technique (see Chapter 1); most make use of rotating grinding surfaces to reduce the requirement for manual effort, either lowering the specimen vertically against a moving horizontal surface (e.g. Sutton et al. 2001a), or pressing it horizontally against a lapping surface. While these devices vary in detail, in the degree of automation supported, and in the scale of specimen for which they are designed, the best quote a grinding-increment precision of around 10 µm; this figure has not improved over the entire 100+ year history of the technique. Details of one particular grinding methodology are given in the case-study in Section 2.2.4.1.

Grinding-based tomography is maximally destructive – unlike slicing or sawing, all material is destroyed. It can also be slow, especially for larger specimens where substantial amounts of material need to be ground away. It does, however, have two key advantages. Firstly, it allows very fine and precisely controlled increments, so can produce far higher resolution than can serial sawing. Secondly, the grinding process, when carried out with appropriate abrasives, doubles as a polishing process, resulting in surfaces well-suited for high-fidelity imaging.

2.2.1.2   Sawing

Sawing involves the exposure of surfaces by sawing through the specimen. While partial saw cuts followed by manual breakage are in theory possible to minimize destruction of material, this approach is fraught with complications for reconstruction and risks uncontrolled breakage; it is not recommended. A saw cut exposes two surfaces, one each side of the **kerf** (the removed slice of material); serial sawing hence produces inconsistent (alternating) spacing between tomograms, half of which need to be mirrored prior to reconstruction as photographs will have been taken in opposite directions (Figure 2.1). Most reconstruction methodologies can compensate for this, although such compensation may not be straightforward, and alternating spacing may produce undesirable **artefacts**. Any type of saw capable of cutting rock can in theory be used; saws with finer kerfs are normally preferred as they minimize the loss of material, but coarser kerfs create greater separation between the exposed faces, which may be advantageous (e.g. to reduce the difference between alternate tomogram spacings). Coarse rock saws typically have a

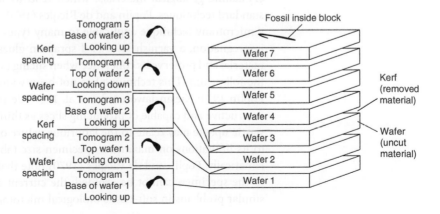

Figure 2.1 Five tomograms from a hypothetical serial-sawing dataset, demonstrating alternating kerf/wafer spacing and mirrored tomogram effect.

kerf of at least 2 mm, and as they normally require a manual feed, they are hard to use precisely – for these reasons we do not recommend them for serial-sawing tomography. Fine-blade low-speed saws with a kerf under 0.4 mm are widely available in laboratories, and are well-suited for serial-sawing work, allowing specimens to be clamped and precisely positioned for cutting. However, their high cut-times can be prohibitive, and they typically cannot cut specimens more than a few centimetres in size. Faster fine-blade saws with similar properties also exist, but are more expensive, can be hazardous to operate, and may have similar size restrictions; see e.g. Joysey and Breimer (1963) and Honjo et al. (1968) for early applications. Wire saws, which cut using a diamond-impregnated wire or a metal wire combined with abrasive slurry, are another option; these are capable of cutting large specimens (they are not restricted by blade-size), and can have finer kerfs than blade saws; values of 0.2 mm may be achievable with industrial machines (Kao et al. 1997), and specialist ultra-slow devices have quoted values as low as 0.1 mm (Fosse et al. 1974). Serial sawing has never attained the popularity of serial grinding, the principle impediment to adoption being the relatively slow cut-times of many saws, which can often approach or exceed an hour. Sawing is also not capable of the fine inter-tomogram resolution achievable by grinding methods, being limited by saw kerf and the physical strength of wafers (slices of inter-kerf material), and does not produce surfaces that are as well-polished. For large specimens not amenable to non-destructive tomography, it may nonetheless be the only practical means of tomographic study.

2.2.1.3 Slicing

Slicing involves serially removing and retaining fine slices of material with a microtome. This technique is widely used in biology, where soft tissues hardened with paraffin (and decalcified hard-tissues) can be sliced in thicknesses as fine as 50 nm, although thicknesses measured in micrometers are more common (see e.g. Bancroft and Gamble 2008). This ultra-fine tomogram spacing is at the top end of what can be attained with any other technique discussed in this book, but is unfortunately not achievable for crystalline geological materials, which tend to shatter when sliced using standard techniques. Poplin and de Ricqles (1970) have, however, described a microtomy technique applicable to many types of fossils, in which resin impregnation, a varnish layer and spray-on glues are used to consolidate material and preserve its coherence when slicing (see case study discussed in the following). This technique has not been widely adopted (although see Poplin 1974; Kielan-Jaworowska et al. 1986) despite being only minimally destructive and capable of producing slices as thin as 15 μm. Lack of enthusiastic uptake may relate to the laborious nature of the method, or perhaps to its limitations on maximum specimen size (about 30 mm). More problematically, Poplin and de Ricqles (1970) note that distortions and damage to the specimen cannot be avoided; the current authors are familiar with similar problems in soft-tissue biological microtomy. These distortions are likely to render satisfactory reconstructions difficult for the same reasons as peel-based datasets are difficult (see Section 2.2.2.3), and thus we do not recommend the approach as a basis for the production of virtual specimens.

Modern high-precision ultramicrotomy can also be carried out within a scanning electron microscope (SEM), the microtome replacing the standard sample stage, and the SEM providing imaging (see e.g. Reingruber et al. 2011). We are only aware of one such tomograph; the Gatan 3 view 2 (www. gatan.com). This can create a **registered tomographic dataset** of up to 600 μm depth with a tomogram spacing from 5 to 200 nm; the process is fully automated. It is designed for resin-mounted biological specimens, but can be used on resin-mounted macerated fossils or those preserved in soft rocks such as shales. As with all microtomy, distortions remain a problem. Rock heterogeneity in particular may cause these; quartz grains, for example, can be pushed through the surrounding rock by the blade. SEM microtomy has yet to be used in palaeontological research, but for very small and suitably soft specimens it may rival FIB tomography (Section 2.3).

2.2.2 Approaches to Imaging

Generation of tomogram images that can be used as a basis for reconstructing a virtual specimen (see Chapter 5) is, in all forms of physical-optical tomography, achieved by photography or manual tracing. This may be undertaken directly from the exposed surface (Sections 2.2.2.1 and 2.2.1.2) or from acetate peels taken of the surface (Section 2.2.1.3).

2.2.2.1 Photography

Direct photography of exposed surfaces is the simplest and most widely applicable approach; photographs can be used directly as the raw data for a reconstruction, although registration (alignment) will normally be required before this is possible (see Chapter 5). Photographs can be taken using any photographic set-up, the scale of the specimen being the most obvious control on the method used; for smaller specimens, either a microscope-mounted camera or bellows-based macrophotographic set-up (see Siveter, Derek J. 1990) may be required. Digital photography is strongly recommended – while film photography is theoretically capable of higher resolution, the practical control on resolution is the quality of the camera optics not the recording medium. Crucially, digital photography allows immediate inspection and quality control of the image, which can be retaken if flawed in any way. For serial-grinding tomography in particular this is vitally important, as mistakes are otherwise unlikely to be noticed until after the tomographic surface concerned has been destroyed. Digital cameras that are directly connected to a computer are preferred over those that store images in on memory card in the camera, as these enable instant inspection (on a large screen) and backup of data. In general, the highest resolution available should be used, with the proviso that resolutions beyond the capabilities of the camera optics offer no benefit. While resolution may need to be reduced for reconstruction (see Chapter 5), this is easily accomplished at a later stage, and it is preferable to have the most data-rich image capture attainable, even if those data are not all used for a particular reconstruction. Desktop document scanners may provide a viable alternative to digital cameras; these are typically capable of scanning at at least 1200 dpi, which equates to about one pixel per 20 μm, and are a relatively cheap way of photographing planar objects at high resolutions and under consistent lighting conditions. Image quality may not, however, match up to that of a good photographic set-up, and wetting (see in the following) is unlikely to be possible.

Prior to photography, surfaces will need to be washed clean of any abrasive powder or slurry. Photography itself is normally best performed through a liquid layer; water, glycerine, xylene or alcohol can be used (Siveter, Derek J. 1990). This reduces the contrast in refractive index at the tomographic surface (these liquids having a higher refractive index than air), which can greatly improve contrast between the specimen and the matrix in which it is embedded. The liquid layer can take the form of a water or other liquid bath, or simply a 'wetting'. Care must be taken in the former case to avoid disturbances such as ripples or waves, and in the latter case to avoid reflections or distortions from any meniscus. In serial-grinding tomography, liquid-layer photography is normally only available when the specimen can be removed from the tomograph during the imaging part of the workflow, unless a tomograph is used where the specimen is held with the tomographic plane horizontal and the exposed surface upwards (e.g. the machine described by Croft 1953). Liquid-layer photography is always available for sawed wafers; these may also require polishing prior to imaging to achieve good contrast.

Reconstruction methodologies require that tomograms are 'independent' (see Table 5.1), that is, that all information within a tomographic image comes from a single plane. Direct photography of translucent specimens may violate this requirement, and is hence not recommended. Independence problems can also arise from rough surface topographies; see Section 2.2.3.4 for one solution.

2.2.2.2 Tracing

Tracing of structures is an alternative to direct use of photographs; historically, this is the form in which most published tomographic datasets were presented. Here, the researcher produces interpretative tracings of the structures which he or she is interested in reconstructing, and uses these rather than the original images as the basis of any reconstruction. Tracing can be made directly from the surface using either a camera lucida or direct overlay of translucent tracing media. Alternatively they can be made indirectly, from photographs of the surface itself, photographs of acetate peels, or direct from acetate peels using a camera lucida. Figure 2.2 provides an example of tracings of the latter type. If the tracings are to be used as the basis for a virtual reconstruction, they need to be digitized; they can either be made directly on a computer (e.g. by tracing over an image using an

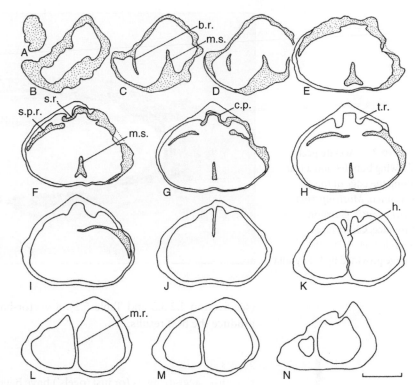

Figure 2.2 Traced tomograms of the brachiopod *Stentorina sagittata*. These were made by tracing direct from acetate peels, using a camera lucida. Scale bar is 0.5 mm. Source: Baker and Wilson (1999, Fig. 1). Reproduced with permission of the Paleontological Association.

appropriate graphics package), or can be drawn manually and then scanned. It should be noted that traced tomograms may lend themselves better to vector-surfacing reconstruction techniques (see Section 5.2.3) than volume-based ones such as isosurfacing (Section 5.2.4.2), and hence it may be advisable to perform tracing using a vector graphics application capable of exporting shapes to the target reconstruction package. Tracing has several important characteristics, which might or might not be perceived as advantages. Firstly, it forces the researcher to interpret 'difficult' areas of the fossil and make a decision as to where the margins of the structure actually lie, rather than leaving that decision to an unintelligent algorithm. Secondly, it acts as a preparation phase, where structures not of interest can be excluded, and discrete structures can be separately traced (e.g. in different colours) with a view to maintaining this separation in any virtual model. Finally, it normally produces tomograms that are clearer and more easily interpreted in themselves than any direct imaging technique; if tomograms are to be presented in a publication instead of (or as well as) a virtual model, tracing may be desirable for this reason alone. It is, however, time-consuming and subjective, potentially introducing worker-bias into the reconstruction. In datasets where the tomogram count is high, we would not normally recommend this approach, preferring to perform preparation and structure identification during the reconstruction phase (see Section 5.2.4.3). However, where tomogram count is low and the dataset is markedly non-isotropic

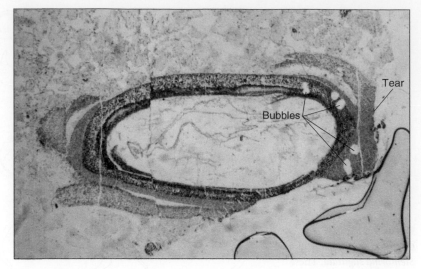

Figure 2.3 Acetate peel showing bubbles and a tear along the edge of the specimen. *Muricosperma guizhouensis*, Guizhou Province, China (see Seyfullah et al. 2010). Image provided by L. Seyfulla.

(see Section 2.2.3.2 and Table 5.2), a vector-based tracing approach may produce the best results.

2.2.2.3   Peels

Cellulose acetate peels (or just 'peels') have been widely used as a means of producing a permanent record of fossil surfaces, for both physical-optical tomography and other purposes. Peels cannot be used directly as the basis for a virtual reconstruction; they must first be either photographed or traced (see Section 2.2.2.2) to produce digital images. Peeling is thus not an alternative to the imaging techniques described earlier, but an extra intermediate stage in the imaging workflow. Details of the method for peel preparation vary and are beyond the scope of this book (see e.g. Galtier and Phillips 1999 for a full treatment), but put simply the surface is briefly etched with weak acid, dried, flooded with acetone, and then a cellulose acetate sheet is applied. The sheet becomes impregnated with the surface materials; when dry it is peeled off, and normally placed between glass plates for storage. While time-consuming to prepare, peels are attractive as they preserve not simply an image but a physical mineralogical record of the surface. They can also be augmented through the application of various chemical stains to the surface, allowing the differentiation through colour of, for instance, calcite from aragonite, or ferrous from non-ferrous calcite. There are, however, some serious problems with the use of peels as a basis for virtual reconstruction (see Figure 2.3). Firstly, peels are prone to 'wrinkling'; even where this effect is apparently minor on visual inspection, it can heavily degrade the quality of any reconstruction by interfering with the positional continuity of structures between tomograms. Peels are also prone to bubbles, which cause areas of information loss with distortions around them, and to tearing, which leads to the displacement of large areas of information. Slight stretching artefacts relating to the direction of pull when peeling can also be present. Finally, consistency of contrast is very difficult to achieve, especially where staining agents are used. None of these artefacts are easily corrected for, and even expertly prepared peel-based datasets are not immune to such problems. The authors' experience with reconstruction of pre-existing peel-based tomographic datasets, many of which also lack fiduciary structures (see Section 2.2.3.3), is that virtual models of acceptable quality are often very difficult to produce. For these reasons we strongly recommend that peels are not used for physical-optical tomography; direct imaging of surfaces is greatly preferable, and will pro-duce far more satisfactory reconstructions. Note that if a peel record is required, there is no reason why peeling cannot be performed *after* photography of a surface.

2.2.3   Other Considerations for Methodology

When implementing a physical-optical tomography methodology intended as the basis for a virtual reconstruction, there are many design issues that need consideration in addition to the selection of a surface exposure and imaging method (see the preceding). These are discussed subsequently, in no particular order.

2.2.3.1   Consistency of Images

For reconstruction purposes images should be as consistent as possible. Consistency of lighting is important – lights should be consistent in bright-ness and in position relative to specimens for all tomograms. All digital images should be taken at the same magnification and resolution, with the same camera, and through the same lenses (as lens distortions can differ). As far as possible, all images should be taken at the same orientation, and with the specimen positioned at least approximately in the same place in the visual field each time. A lack of consistency in most of these regards can be corrected for during the reconstruction phase, but it is preferable to avoid the need for such corrections. Additionally, if precise consistency of position and orientation can be achieved here, the registration phase of reconstruc-tion (see Section 5.2.2) can be skipped, although this is extremely difficult to achieve in practice.

2.2.3.2   Tomogram Frequency

Ideally, spacing between tomograms should be as low as the tomograph used is capable of, to maximize data retention. However, practical consid-erations of time availability may prompt a higher spacing. Isosurface and other **volume-based reconstruction** methodologies (see Section 5.2.4) normally require consistent slice spacing for the entire dataset; although some software is able to correct for inconsistencies, it is preferable to maintain the same spacing throughout. In some simple grinding tomography set-ups it is not possible to precisely control the amount of material removed each time; in these cases, it is vital that the thickness removed is at least measured, otherwise correction for inconsistency will be forced to rely on guesswork. The slice spacing used may also be influenced by the requirement of isosurface-based reconstruction techniques for near isotropy in the dataset – that is, that the spacing between slices should be not be more than a factor of 3 or 4 different to the distance between pixels in the tomogram images, and ideally should be the same (see Chapter 5). This can be achieved by **downsampling** the tomograms prior to reconstruction, but as this discards data it is preferable to collect tomograms at a high enough frequency to avoid or minimize downsampling.

2.2.3.3   Fiduciary Markings

Almost all physical-optical methodologies (though see Section 2.2.3.1) require that images are registered (or aligned) prior to reconstruction – Section 5.2.3 details the procedures for this. Registration should also be considered at data-capture time, however, as it becomes far easier and less prone to error if **fiduciary markings** are included in the original images. Fiduciary markings are invariant points or linear features that can be used as a guide for registration; examples include holes drilled through the sample perpendicular to the tomographic plane, other vertical structures (such as pencil leads) set in resin close to the specimen, cut edges perpendicular to the tomographic plane, or other planar structures set close to the specimen such as the walls of a retaining box (see Figure 2.4). Sufficient fiduciary markings should be included to detect rotational, translational, and ideally also scaling inconsistencies between tomograms, for example, at least two holes or three edges (see Figure 2.4d–f). Drilled holes are less destructive, and are normally preferred where the position of the specimen within a sample is uncertain. Cut edges can, however, be more precise, especially for smaller specimens where sufficiently fine drilling is impractical. The placement of fiduciary markings for matrix-embedded specimens where the exact position or shape of the specimen is uncertain is an uneasy compromise. If markings are placed too far from the specimen, then the field of view of each image will need to be large to include them, reducing the resolution covering the specimen itself. If they are placed too close, there is a risk of damaging the specimen by drilling or sawing through it.

Ideally, fiduciary markings should be perpendicular to the tomographic plane, so that they are genuinely invariant in position for all images. In practice, it may be difficult to guarantee that the angle between them and the plane is exactly 90°; this is particularly true for cut edges on small specimens. Where any such discrepancies exist, they will result in 'drift' of the fiduciary markings through the tomographic dataset, and it is important that they are measured in some way so that this drift can be corrected for at reconstruction time. Without such correction, the resulting virtual specimen will display a degree of skew.

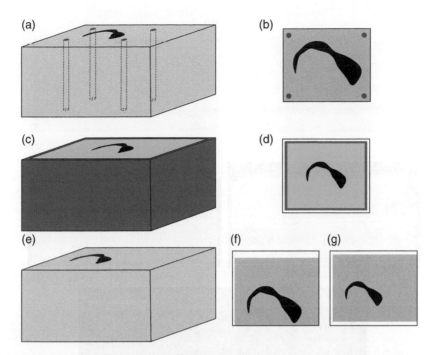

Figure 2.4 Fiduciary markings, used as an aid for registration (alignment) of tomograms. (a) Block with specimen and vertically drilled holes. (b) Tomogram from (a); holes are placed close to specimen, which can thus be large in the field of view. (c) Specimen isolated and set in blue resin within an orange retaining box. (d) Tomogram of (c), with box providing a fiduciary structure. (e) Block with cut edges but no vertical holes. (f) Tomogram of (e) capturing two edges – these structures will control for translation and rotation errors, but not errors in scaling. (g) Tomogram of (e) capturing three edges – these will control for translation, rotation and scaling errors, but specimen is smaller in field of view than in (f).

2.2.3.4 Other Specimen Preparation

In addition to the emplacement of fiduciary structures, other specimen preparation work may need to be carried out prior to data capture. Many tomographs, especially grinding tomographs, have a maximum physical size that they can accommodate, and a maximum depth of grinding. Camera maximum field of view may provide another limit on specimen size. Blocks containing specimens may need to be trimmed (using a saw) to comply with these strictures, and where specimen depth is more than the maximum grinding depth, the specimen may need to be cut into more than one piece before processing. This sawing is necessarily destructive, and it should be carried out with a fine-kerf saw (see Section 2.2.1.2) to minimize data loss if there is any risk of it intersecting the specimen.

Specimens that consist of a part and counterpart can be reconstructed in two ways; either part or counterpart can be glued together prior to reconstruction, or they can be reconstructed separately and reunited digitally after reconstruction. The authors' experience is that the latter normally produces more satisfactory results, although the former may be quicker. If part and counterpart are glued together, it is important to ensure a precise positioning.

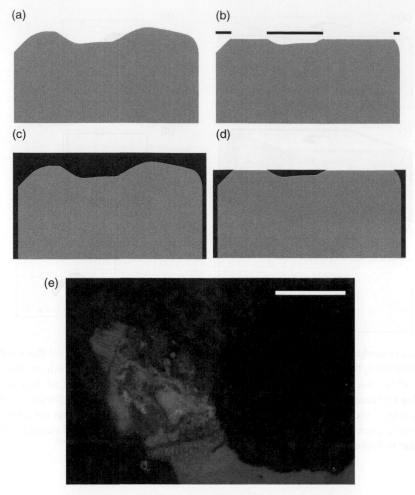

Figure 2.5 The use of opaque resin to hide structures outside the tomographic plane arising from rough surfaces. (a) Side view of specimen with a rough top surface. (b) Specimen after removal of some material – the regions below the black bars will be visible in any photograph taken from above, but are not in the tomographic plane. (c) Side view of same specimen set in opaque blue resin. (d) Specimen set in resin after removal of some material – all specimen below the tomographic plane is hidden by resin. (e) Photographic tomogram of the holotype of *Kulindroplax perissokomos* Sutton et al. (2012b) showing specimen partially obscured by resin as in (d); scale bar is 5 mm.

Finally, there are also several reasons why it may be helpful to set the specimen in resin prior to tomography, particularly if the resin is rendered opaque and coloured by the addition of a dye. Firstly, isolated specimens may be difficult to mount on a tomograph; setting these in resin may assist, as the resin block can then be cut into a more manageable shape, and/or used to hold fiduciary structures. Secondly, setting in dyed resin may make fiduciary edges clearer in tomograms – for instance, a red specimen could be set in a blue dye. Finally, and most importantly, an opaque resin is useful where a specimen has a rough upper and lower surface – see Figure 2.5.

If no resin is used, photographs of this surface will include information below the tomographic plane, violating the reconstruction requirement of slice independence (see Table 5.1). An opaque resin will obscure this out-of-plane material; although the resin will still need to be removed from the tomogram prior to reconstruction.

2.2.3.5 Multiple Specimens

One way to increase the efficiency of a physical-optical workflow is to mount multiple specimens on a tomograph, potentially greatly decreasing the grinding or sawing time per specimen, and also increasing the efficiency of other operations (e.g. washing). Photography, if applicable, should normally still be carried out on a per-specimen basis, as attempts to capture multiple specimens with a single photograph will reduce the resolution covering each specimen. Nonetheless, photography will still be more efficient on multi-specimen workflows, as operations such as focusing will only need to be carried out once per cycle.

2.2.4 Case Studies of Methodology

To provide exemplars of particular methodological workflows, we detail three different physical-optical methodologies; these have not been selected with a view to firmly recommending any particular methodology, as details will always depend on the samples to be analyzed and the equipment available. The first of these is the methodology with which the authors have most familiarity, and is hence provided in the most detail.

2.2.4.1 Grinding – the Herefordshire Lagerstätte

The Silurian Herefordshire Lagerstätte (Briggs et al. 1996) preserves a diverse fossil fauna of soft-bodied invertebrates in three-dimensional form. The fossils are for the most part preserved as sparry-calcite crystals within a carbonate-cemented matrix of volcanic ash, and vary in size from a few millimetres to a few centimetres. They have not proved amenable to non-destructive imaging of any type, and have hence been reconstructed using a grinding-based physical-optical method over a series of publications (e.g. Sutton et al 2001a; Siveter, Derek J. et al. 2003, 2004; Sutton et al. 2012b). Optical imaging of physically exposed surfaces is particularly applicable for this material as optical contrast between specimen and matrix is high (see Figure 2.6). The methodology uses a Buehler slide holder (Figure 2.7d–f) as a simple but precise hand-held tomograph; it was described by Sutton et al. (2001b), with some supplementary details added by Sutton et al. (2012a).

Specimens selected for reconstruction were first cut to size. This was accomplished using a coarse rock saw, where it was possible to be confident that the specimen would not be intersected; where there was any risk of the saw cut damaging the specimen a Buehler IsoMet low-speed saw with a

Figure 2.6 Section through the polychaete worm *Kenostrychus clementis* Sutton et al. (2001c) from the Herefordshire Lagerstätte, showing clear optical contrast between fossil (dark) and matrix (light); scale bar is 1 mm.

0.2 mm blade and a kerf of approximately 0.3 mm, was used. Maximum sample size was primarily controlled by the maximum depth of the slide holder – about 10 mm – and the maximum camera field of view – about 30 mm. Where specimens were larger than this maximum, they were cut into smaller pieces, using the fine Buehler saw described earlier. The edge faces produced by sawing-to-size doubled as fiduciary markings, and were hence cut perpendicular to each other and to the direction of grinding, and as close to the specimen as possible (Figure 2.7b). Side faces were polished to remove roughness from sawing. Specimens were then set within a cylinder of blue-dyed resin (Figure 2.7c). This was accomplished by placing them face-down at the base of a cylindrical mould with a diameter of around 30 mm, then carefully pouring dyed resin over them and leaving them to set. Note that while Figure 2.7 illustrates a grinding run performed on a single fossil, multiple specimens (2–8, depending on size) were normally mounted within one resin cylinder; a single grinding run normally involved three resin cylinders mounted adjacent to each other. The exact position of each specimen within the cylinder was recorded with notes and photographs prior to the application of resin, so the identity of specimens and the correct photographic orientation could be deduced once they began to emerge from resin during the grinding run. The resin used was Buehler EpoxiCure & Hardener (20-8130-032/ 20-8132-008), and the dye was Buehler Epoblue powder (11 10 68); blue was chosen to enhance contrast with the (normally) orange matrix.

Prior to grinding, the meniscus and any superfluous thickness of resin were removed from the top of the resin cylinder using rock saws and lapping wheels, so that the depth of the cylinder fell within the maximum grindable depth of the slide holder, and the top was a flat surface suitable for mounting. Care was taken that this surface was perpendicular to the cylinder walls.

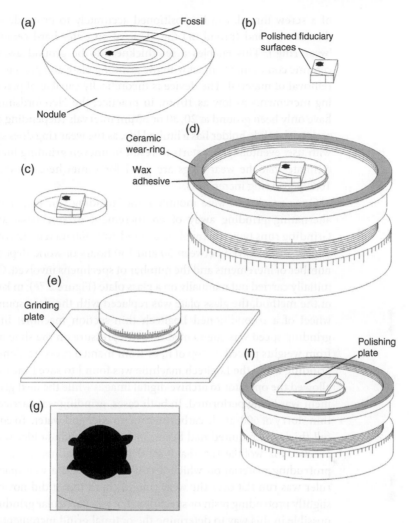

Figure 2.7 Serial-grinding methodology used for the Herefordshire Lagerstätte fossils. (a) Specimen in split nodule prior to processing. (b) Specimen cut to size for grinding, with polished faces as fiduciary markings. (c) Cut specimen emplaced in resin disk. (d) Resin disk mounted on slide holder. (e) Slide holder placed on glass plate for manual grinding – note that in later iterations of this method the glass plate was replaced with a thin-section lapping machine. (f) Specimen being polished prior to imaging – note that in later iterations of this method this step was typically discarded. (g) Photographic tomogram of specimen capturing two fiduciary edges. *Source*: Sutton et al. (2001b, Fig. 2). Reproduced with permission of the Palaeontological Association.

Where multiple cylinders were mounted, they were cut to as near the same depth as possible. Resin cylinders were then attached to the slide holder by placing it on a hotplate at around 80°C, melting thermal mounting wax onto the slide-holder surface, emplacing the cylinders, and then quenching in cold water.

The Buehler 'slide holder' or 'specimen holder' (part numbers 60-8113 or 60-8103, Figure 2.7 d–f) consists of an inner metal cylinder which, by means of a screw thread, can be positioned accurately to protrude any required distance beyond (proud of) the plane of an outer hard ceramic disc (the 'wear ring'). This enables a set thickness to be ground away before the ceramic comes into contact with the grinding plate and prevents any further removal of material. The device is theoretically capable of precisely removing increments as low as 10 μm; in practice, the Herefordshire specimens have only been ground at 20, 30 or 50 μm intervals (depending on specimen scale). The slide holder has a limited life, as the wear ring does slowly abrade with use; eventually this starts to result in uneven grinding increments. We estimate that the wear rings are good for somewhere between 5000 and 10,000 grinding increments.

Once specimens were mounted, the 'grinding run' began, consisting of alternating grinding away of an increment of thickness, and imaging. Grinding runs typically involved around 300–400 increments of either 20 or 30 μm, and involved between 50 and 150 hours of work, depending on the number of increments and the number of specimens involved. Grinding was initially carried out manually on a glass plate (Figure 2.7e); in later iterations of the method, the glass plate was replaced with the slow-spinning lapping wheel of a reconditioned Logitech thin-section machine. In both cases, grinding speed was augmented by extra pressure on the slide holder, either from weights placed on top of it, or from manual pressure. Semi-automated grinding using the Logitech machine was found to save time primarily as it allowed the operator to archive digital images while the next grinding increment was being performed. In both cases, grinding was carried out using a thin slurry of 600-grade carborundum powder and water. To ensure that the full thickness required had been removed, the slide holder was repeatedly checked for 'wobble' (i.e. that it sat flat on the grinding surface without any protruding material on which it could pivot), and a well-machined metal ruler was run flat over the wear ring to check that it did not 'catch' on any slightly protruding resin or specimen. For any particular grinding run it was possible in this way to determine the optimal grind increment time by trial and error; typically this was 5–10 minutes.

After grinding, the slide holder was inverted and placed on its base, the surface washed with a directable-jet water bottle, and polished for a few seconds with a small flat plate coated in fine-grade aluminium oxide powder and water (Figure 2.7f). This final polishing was initially thought to improve image quality, but was later dropped from the process when it became clear that any such improvement was negligible. Where polishing was performed, the surface was then washed again. The slide holder was then placed under a Leica MZ8 binocular microscope with the specimen(s) still wet, and photographs taken through the remaining film of water (see Section 2.2.2.1 for rationale). Photography was accomplished using several different cameras over the 11+ years for which this methodology has been used; lately in use is a Leica DFC420, which captures 24-bit colour images at a resolution of 2592 × 1944 pixels direct to a laptop computer. All cameras used have been mounted on a photo-tube attached to the MZ8 microscope, with a microscope-mounted ring-light used for consistent lighting. Maximum horizontal field of view using standard lenses is around 30 mm; this could be increased to around 50 mm by using appropriate objective lenses, but this was found to result in minor but significant distortion near the edges of the image. Each image was manually inspected for quality – the main problems encountered were pieces of floating material obscuring the specimen, focus errors, or reflection problems relating to overly thick or thin layers of water. Where necessary, images were re-taken. Once deemed satisfactory, they were filed as part of a sequential set in a lossless file-format (normally .bmp); as noted earlier, saving and naming of files was normally undertaken in parallel with the next grinding iteration. For grinding runs involving multiple specimens, focus was normally checked once every two imaging iterations, and the only setting adjusted for each specimen was the magnification, the optimal setting for which typically varied between specimens. Careful records of magnifications used were kept, and care was also taken to precisely reset magnification each time. Total imaging time (including washing) was normally 5–10 minutes, depending on the number of specimens; total grinding increment times thus varied from around 10–20 minutes.

The reconstruction methodology used to produce virtual reconstructions from these datasets is discussed in Section 5.6.1.

2.2.4.2 A Fast-Saw Methodology

Maisey (1975) described a serial-sawing technique using a Capco Q-35 automatic high-precision cutting machine; this methodology was first employed by Joysey and Breimer (1963) to study a Carboniferous blastoid. While the method is not recent, it is not dated in any important regards. Furthermore, one of us (MDS) has experience with a similar saw, and can confirm its efficacy for cutting fossil material. Details of the method from Maisey (1975) are reproduced here.

The Capco Q-35 is an inner-diameter saw, that is, a circular blade with a large central circular hole, where the cutting surface is the inner rather than outer edge of the blade. It uses an annular stainless-steel blade impregnated with diamond on the cutting surface. The blade is mounted within a steel drum, and rotates at around 2400 rpm. The saw has a kerf of approximately 0.17 mm, and can accommodate samples up to 30 mm high, 40 mm wide and 50 mm long. Cutting takes place within a spray of oil coolant.

Prior to cutting, Maisey (1975) recommended embedding in resin to ensure cohesion during cutting; while he notes that certain materials may have enough inherent strength for this not to be required. However, embedding sometimes brings other advantages (see Sections 2.2.3.3 and 2.2.3.4), may make the specimen easier to mount, and can render wafers easier to handle, obviating the need to touch the specimen itself. Maisey noted that the embedding medium needs to react well to sawing (most critically it needs not to melt at the point of contact), and be of a sufficiently low density so as not to greatly increase sawing time. He recommended Trylon polyester EM

301 copolymer resin, partly for its transparency, which facilitated orientation of specimens prior to cutting (although see Section 2.2.3.4). Once embedded, the resin block was cut into a rectangular prism. The resin block was soldered to a ceramic bar using a high-temperature sealing wax, and the ceramic bar in turn attached using wax to a brass bar, which was clamped onto the saw's moving table. The ceramic bar was present as saw cuts often extended below the specimen into the substrate; the saw was set up so that the blade could cut into the ceramic where necessary, but not into the brass bar below.

When cutting, the saw platform was pulled outwards onto the blade by means of weights, and the rate of movement manually controlled. Cutting time for one cut varied from a few to over 45 minutes, depending on specimen size and hardness. The Capco saw was capable of operating in an automated mode, where after each cut the specimen was withdrawn from the blade, advanced a set interval and then cut again. The minimum possible cutting interval was 0.25 mm, resulting in wafers 0.08 mm thick; Maisey (1975) recommended a minimum cutting interval of 0.3 mm for fossil material, resulting in stronger wafers 0.13 mm thick.

After all cuts had been made, wafers were removed from the block; where the resin was not fully cut through, this was done using fine scissors or a scalpel. Oil was removed from sections using acetone, and up to 30 were then mounted on a single slide. Prior to imaging, wafer surfaces were polished using carborundum powder; our experience with similar saws, however, is that such polishing is not always required where reflected light photography is to be used. Maisey did not specify any image-capture methodology, but the wafers produced by his technique could easily be imaged, for example, using the photographic set-up described in Section 2.2.4.1.

2.2.4.3 A Microtomy (Slicing) Methodology

The methodology described by Poplin and de Ricqles (1970) is the only microtomy technique of which we are aware that has been used for mineralized palaeontological specimens (although see Section 2.2.1.3); we reproduce details of it here for completeness and for historical interest, and despite the caveat that we consider it unsuitable as a basis for virtual reconstructions (see Section 2.2.1.3), we note that it has been used as the basis for wax models (Kielan-Jaworowska et al. 1986). Quantitative details of the typical time-per-slice of this method are not available, although the original authors describe it as relatively rapid. They quoted a maximum specimen diameter of 20–30 mm, limited by the size of the microtome blade.

Prior to microtomy, specimens were impregnated with a relatively fluid resin – Epikote 828 Shell, DETA (diethylen triamine) hardener, and Araldite DY 021 in the ratio 9:1:3, respectively, by weight. Impregnation was considered essential to ensure coherence of porous material when slicing – for material with negligible porosity, it may be skipped. After impregnation, the specimen was set in a rectangular resin block (Figure 2.8a, resin block A) with parallel sides perpendicular to the intended tomographic plane and as close to the specimen as possible, to act as fiduciary markings. For this

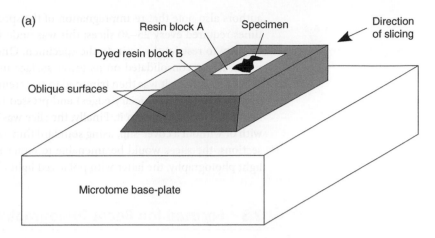

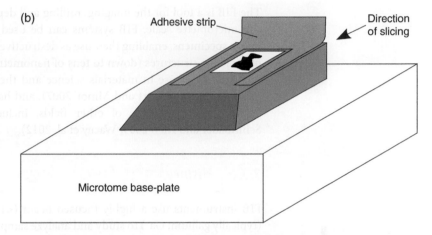

Figure 2.8 Parts of the procedure of Poplin and de Ricqles (1970). (a) Specimen mounted inside two resin blocks on microtome base-plate. (b) Use of adhesive strip to stabilize slices, cut with gap for specimen. Modified from Poplin and de Ricqles (1970, Fig. 1). Reproduced with permission of Wiley-Blackwell.

second resin, the original authors recommended 'Vestopal H', but noted that other resins were viable, so long as the hardness after setting was such that a scalpel could easily remove tiny shavings. They then set this small block into a much larger block of the same resin (Figure 2.8a, resin block B), but this time incorporating a dye so that the fiduciary edges of the small block were visible. This block was cut with oblique edges away from the direction in which the cutting blade was to come from; this oblique angle functioned to prevent splintering. The upper surface was polished, and the resin block was then clamped or otherwise mounted onto a Jung microtome ('model K'), which was equipped with a tungsten carbide steel blade.

Once mounted, the serial microtomy was begun, with thicknesses varying from 15 to 30 μm. For each slice, an adhesive paper cut with a slot for the specimen (Figure 2.8b) was placed atop the resin block prior to slicing to help hold the material together. A consolidating flexible varnish on top of each slice was also required for less porous material that had not been well-impregnated by the fluid resin described earlier; where required this was applied as two or three layers of Plexigum, diluted in acetone. The original authors also note that re-impregnation of the specimen by resin was sometimes required every 25–30 slices; this was undertaken *in situ*, using a wall of putty to restrict the resin to the specimen. Once a viable slice had been made, it was consolidated on its lower surface using a light spray-on glue such as hairspray. It was then trimmed to size (removing the adhesive paper, but preserving the fiduciary edges) and pressed between glass plates for at least 15 minutes to flatten it. Finally, the slice was mounted on a glass slide, with or without a cover-slip, using standard thin-section techniques. As thin sections, the slices would be amenable to either reflected- or transmitted-light photography, the latter with polarized light if necessary.

2.3   Focused Ion Beam Tomography

The FIB is a tool for the imaging, milling and deposition of material at the sub-micrometre scale. FIB systems can be used to sequentially mill and image specimens, enabling their use as destructive tomographs for the study of very fine structures (down to tens of nanometres in size). The FIB has a long history of use in materials science and the semiconductor industry (Phaneuf 1999; Volkert and Minor 2007), and has recently started to find applications in a range of other fields, including palaeontology (e.g. Schiffbauer and Xiao 2009; Wacey et al. 2012).

2.3.1   History

FIB instruments use a highly focused beam (<1 μm in diameter) of ions (typically gallium; Ga⁺) to study and analyze samples. The first such systems were developed in the mid-1970s and relied on gas field-ionization sources (Levi-Setti 1974; Escovitz et al. 1975; Orloff and Swanson 1975). The subsequent development of systems with liquid–metal ion sources in the late 1970s and early 1980s (Seliger et al. 1979; Swanson 1983), capable of higher imaging resolutions, led to the widespread adoption of the technique in the semiconductor industry; commercial systems became available in the late 1980s (Melngailis 1987; Orloff 1993). Since the 1990s, the FIB has been widely used in materials science, particularly for the preparation of samples for transmission electron microscopy (TEM) (e.g. Overwijk et al. 1993; Giannuzzi and Stevie 1999), but also for *in situ* sectioning and imaging (Phaneuf 1999) and, more recently, tomography (Kubis et al. 2004) and micromachining (Langford et al. 2007).

2.3.2   Principles and Practicalities

The majority of modern FIB systems utilize a gallium liquid–metal ion source; a strong electric field is applied to the source in order to generate an

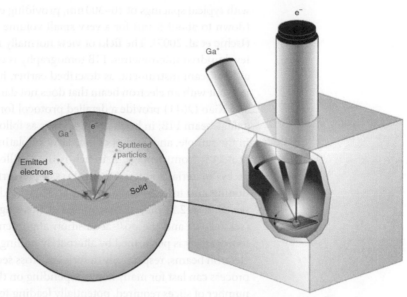

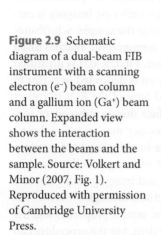

Figure 2.9 Schematic diagram of a dual-beam FIB instrument with a scanning electron (e⁻) beam column and a gallium ion (Ga⁺) beam column. Expanded view shows the interaction between the beams and the sample. Source: Volkert and Minor (2007, Fig. 1). Reproduced with permission of Cambridge University Press.

ion beam, which is then accelerated down the FIB column until it reaches the sample. When the beam strikes the sample, it removes (sputters) a very fine amount of material (nanometres; the exact amount is governed by the amount of time the beam is applied for, and will also depend on the sample properties) from the surface, enabling precise milling. The FIB can also be used to add material to the sample surface via a gas-injection system (MoberleyChan et al. 2007). In addition, a small **region of interest** (ROI) can be imaged using the ion beam; secondary electrons and secondary ions are generated by the sample as the beam scans across it, and these can be detected to produce an image – however, even at low currents, the surface of the sample is damaged by sputtering during this process. This led to the development of dual-beam instruments (Krueger 1999), which combine a scanning electron-beam column and an ion-beam column (Figure 2.9). These systems are capable of high-magnification imaging of structures at nanometre-scale resolutions using the non-destructive electron beam, yielding images based on the detection of secondary electrons (like those produced by a scanning electron microscope) while minimizing damage to the specimen. Moreover, they allow for simultaneous electron imaging and ion milling, making them flexible instruments for observing and modifying samples. FIB machines can be equipped with packages for energy dispersive X-ray spectroscopy (EDS), electron backscattered diffraction (EBSD), and secondary ion mass spectrometry (SIMS), facilitating additional chemical and crystallographic studies, but these modes noticeably increase imaging times.

FIB tomography involves the sequential milling and imaging of parallel surfaces in a sample. This destructive process generates a stack of tomograms with typical spacings of 10–300 nm, providing exceptionally high resolution (down to about 5 nm) for a very small volume (e.g. 1–100 μm³) of material (Uchic et al. 2007). The field of view normally ranges from about 1 μm to a few hundred micrometres. FIB tomography is conventionally performed on a dual-beam instrument, as described earlier, in which case imaging is carried out with an electron beam that does not damage the sample. Schiffbauer and Xiao (2011) provide a detailed protocol for tomographic analysis using a dual-beam FIB; in brief, the process is as follows. First, a ROI is identified in the sample, and a thin (<1 μm) layer of platinum is deposited over this to protect it from damage. A trench is then milled around the ROI to collect excess material sputtered from the sample surface during subsequent sectioning and to reduce shadowing artefacts. Fiduciary markings (e.g. holes; see Section 2.2) may also be milled at this stage to aid automated routines (see below) and the post-acquisition alignment of images. Following this, tomography is performed by alternately milling and imaging (using ion and electron beams, respectively) a series of cross sections through the ROI. This process can last for many hours depending on the sample properties and the number of slices required, potentially leading to drift, but the procedure can be expedited and drift corrected through the implementation of automated control scripts (Uchic et al. 2007). Once data capture is complete, the resulting tomograms must be registered (aligned), often by hand. Subsequently, the dataset can be reconstructed as a three-dimensional model in a similar manner to images gathered by other tomographic techniques (digital reconstruction is discussed in detail in Chapter 5).

2.3.3 Examples in Palaeontology

In the last decade, FIB has been commonly applied in palaeontology for the purposes of sample preparation (e.g. Wirth 2004; Kempe et al. 2005; Bernard et al. 2007, 2009; Wacey et al. 2011; Galvez et al. 2012), but only very rarely in tomographic analyses. Schiffbauer and Xiao (2009, 2011) used FIB tomography to study acritarchs – small, organic-walled microfossils – from the Ruyang Group (~1400–1300 Ma) of North China. They produced cross sections at spacings of 100 or 250 nm, depending on the width of the region of interest (8.2 and 15.4 μm, respectively). These revealed ultrastructures in the fossils that are similar to those imaged in other Precambrian acritarchs using TEM, as well as formerly unknown features. Wacey et al. (2012) looked at more ancient microfossils (2 to >10 μm in size) from the Strelley Pool Formation (~3400 Ma) of Western Australia and the Gunflint Formation (~1900 Ma) of Canada. They performed FIB tomography on 35 μm wide sections with typical slice spacings of 200 nm. The resulting tomograms (Figure 2.10) allowed them to describe a range of microstructures that could be key for identifying microfossils in future FIB-based studies of Precambrian material.

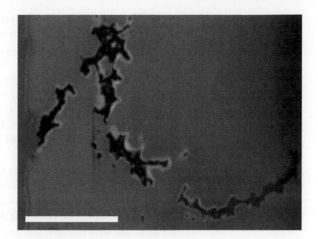

Figure 2.10 FIB tomograms of microstructures from the Strelley Pool Formation of Western Australia. Source: Wacey et al. (2012, Fig. 7). Reproduced with permission of Elsevier.

2.3.4 Summary

Recent studies (Schiffbauer and Xiao 2009, 2011; Wacey et al. 2012) have demonstrated the great potential of FIB tomography for studying microfossil ultrastructure in three dimensions. The nanometre-scale resolution of this technique is around an order of magnitude finer than that which can be achieved using other destructive tomographic techniques. Furthermore, the slices produced by FIB systems contain information about the sample's composition – or chemical/crystallographic structure if EDS, EBSD or SIMS detection is carried out – and thus the method is particularly well-suited for imaging small specimens with low X-ray absorption contrast (not normally amenable to computed tomography). The technique also has several important drawbacks, however, beyond those inherent in all destructive methods. FIB tomography is extremely time-consuming (milling times of up to one hour per section for mineralized samples) and laborious – although automated control scripts (Uchich et al. 2007) could help to alleviate the latter issue. In addition, the ROI is very limited in size (tens of micrometres). Nevertheless, the high resolution and potential to incorporate compositional data render FIB tomography a powerful tool for investigating sub-micrometre structures in abundant (micro)fossil specimens, and it should become more widely used as palaeontologists gain familiarity with the approach.

References

Baker, P.G. & Wilson, M.A. (1999) The first thecideide brachiopod from the Jurassic of North America. *Palaeontology*, **42 (5)**, 887–895.

Bancroft, J.D. & Gamble, M. (2008) *Theory and Practice of Histological Techniques.* Elsevier Health Sciences, Amsterdam.

Bernard, S., Benzerara, K., Beyssac, O., et al. (2007) Exceptional preservation of fossil plant spores in high-pressure metamorphic rocks. *Earth and Planetary Science Letters*, **262 (1–2)**, 257–272.

Bernard, S., Benzerara, K., Beyssac, O., et al. (2009) Ultrastructural and chemical study of modern and fossil sporoderms by scanning transmission X-ray microscopy (STXM). *Review of Palaeobotany and Palynology*, **156 (1–2)**, 248–261.

Briggs, D.E.G., Siveter, David J. & Siveter, Derek J. (1996) Soft-bodied fossils from a Silurian volcaniclastic deposit. *Nature*, **382 (6588)**, 248–250.

Croft, W.N. (1953) A simplified parallel grinding instrument. *Journal of Natural History Series 12*, **6 (72)**, 915–918.

Escovitz, W.H., Fox, T.R. & Levi-Setti, R. (1975) Scanning transmission ion microscope with a field ion source. *Proceedings of the National Academy of Sciences of the United States of America*, **72 (5)**, 1826–1828.

Fosse, G., Röli, J. & Knudsen, H. (1974) A sectioning machine for teeth and other brittle materials. *Acta Odontologica Scandinavica*, **32 (5)**, 299–304.

Galtier, J. & Phillips, T. (1999) The acetate peel technique. In: Jones, T.P. & Rowe, N.P. (eds), *Fossil Plants and Spores*, pp. 67–70. Geological Society, London.

Galvez, M.E., Beyssac, O., Benzerara, K., et al. (2012) Morphological preservation of carbonaceous plant fossils in blueschist metamorphic rocks from New Zealand. *Geobiology*, **10 (2)**, 118–129.

Giannuzzi, L.A. & Stevie, F.A. (1999) A review of focused ion beam milling techniques for TEM sample preparation. *Micron*, **30 (3)**, 197–204.

Honjo, S., Minoura, N., Kumano, S., et al. (1968) Note on serial sectioning of fossil specimen by wheel-saw. *Journal of the Faculty of Science, Hokkaido University. Series 4, Geology and Mineralogy*, **14 (2)**, 175–180.

International Commission on Zoological Nomenclature (ICZN) (1999) *International Code of Zoological Nomenclature* (4th edition). International Trust for Zoological Nomenclature, London.

Joysey, K.A. & Breimer, A. (1963) The anatomical structure and systematic position of *Pentablastus* (Blastoidea) from the Carboniferous of Spain. *Palaeontology*, **6 (3)**, 471–490.

Kao, V., Prasad, J.L. & Bhgavat, M. (1997) Wafer slicing and wire saw manufacturing technology. In: *Proceedings of the 1997 NSF Design and Manufacturing Grantees Conference*, pp. 239–240. National Science Foundation, Seattle.

Kempe, A., Wirth, R., Altermann, W., et al. (2005) Focussed ion beam preparation and in situ nanoscopic study of Precambrian acritarchs. *Precambrian Research*, **140 (1–2)**, 36–54.

Kielan-Jaworowska, Z., Presley, R. & Poplin, C. (1986) The cranial vascular system in taeniolabidoid multituberculate mammals. *Philosophical Transactions of the Royal Society of London. Series B, Biological Sciences*, **313 (1164)**, 525–602.

Krueger, R. (1999) Dual-column (FIB–SEM) wafer applications. *Micron*, **30 (3)**, 221–226.

Kubis, A.J., Shiflet, G.J., Dunn, D.N., et al. (2004) Focused ion beam tomography. *Metallurgical and Materials Transactions A*, **35 (7)**, 1935–1943.

Langford, R.M., Nellen, P.M., Gierak, J., et al. (2007) Focused ion beam micro- and nanoengineering. *MRS Bulletin*, **32 (5)**, 417–423.

Levi-Setti, R. (1974) Proton scanning microscopy: feasibility and promise. In: Johari, O. & Corvin, I. (eds), *Scanning Electron Microscopy*, p. 125. IIT Research Institute, Chicago.

Maisey, J.G. (1975) A serial sectioning technique for fossils and hard tissues. *Curator*, **18 (2)**, 140–147.

Melngailis, J. (1987) Focused ion beam technology and applications. *Journal of Vacuum Science and Technology B*, **5 (2)**, 469–495.

MoberleyChan, W.J., Adams, D.P., Aziz, M.J., et al. (2007) Fundamentals of focused ion beam nanostructural processing: below, at, and above the surface. *MRS Bulletin*, **32 (5)**, 424–432.

Orloff, J.H. & Swanson, L.W. (1975) Study of a field-ionization source for micro-probe applications. *Journal of Vacuum Science and Technology*, **12 (6)**, 1209–1213.

Orloff, J.H. (1993) High-resolution focused ion beams. *Review of Scientific Instruments*, **64 (5)**, 1105–1130.

Overwijk, M.H.F., van den Heuvel, F.C. & Bulle-Lieuwma, C.W.T. (1993) Novel scheme for the preparation of transmission electron microscopy specimens with a focused ion beam. *Journal of Vacuum Science and Technology B*, **11 (6)**, 2021–2024.

Phaneuf, M.W. (1999) Applications of focused ion beam microscopy to materials science specimens. *Micron* **30 (3)**, 277–288.

Poplin, C. & de Ricqles, A.M. (1970) A technique of serial sectioning for the study of undecalcified fossils. *Curator*, **13 (1)**, 7–20.

Poplin, C. (1974) *Étude de quelques Paleoniscides pensylvaniens du Kansas*. Cahiers de paleontologie, C.N.R.S, Paris.

Reingruber, H., Zankel, A., Mayrhofer, P., et al. (2011) Quantitative characterization of microfiltration membranes by 3D reconstruction. *Journal of Membrane Science*, **372 (1–2)**, 66–74.

Schiffbauer, J.D. & Xiao, S. (2009) Novel application of focused ion beam electron microscopy (FIB-EM) in preparation and analysis of microfossil ultrastructures: a new view of complexity in early eukaryotic organisms. *Palaios*, **24 (9)**, 616–626.

Schiffbauer, J.D. & Xiao, S. (2011) Paleobiological applications of focused ion beam electron microscopy (FIB-EM): an ultrastructural approach to the (micro)fossil record. In: Laflamme, M., Schiffbauer, J.D. & Dornbos, S.Q. (eds), *Quantifying the Evolution of Early Life: Numerical Approaches to the Evaluation of Fossils and Ancient Ecosystems*, pp. 321–354. Springer, Dordrecht.

Seyfulla, L.J., Hilton, J., Liang, M.-M., et al. (2010) Resolving the systematic and phylogenetic position of isolated ovules: a case study on a new genus from the Permian of China. *Botanical Journal of the Linnean Society*, **164 (1)**, 84–108.

Seliger, R.L., Ward, J.W. & Kubena, R.L. (1979) A high-intensity scanning ion probe with submicrometer spot size. *Applied Physics Letters*, **34 (5)**, 310–312.

Siveter, David J., Sutton, M.D., Briggs, D.E.G., et al. (2003) An ostracode crustacean with soft parts from the Lower Silurian. *Science*, **302 (5651)**, 1749–1751.

Siveter, Derek J. (1990) Photography. In: Briggs, D.E.G. & Crowther, P.R. (eds), *Palaeobiology: A Synthesis*, pp. 505–508. Blackwell, Oxford.

Siveter, Derek J., Briggs, D.E.G., Sutton, M.D., et al. (2004) A Silurian sea spider. *Nature*, **431 (7011)**, 978–980.

Sutton, M.D. (2008) Tomographic techniques for the study of exceptionally preserved fossils. *Proceedings of the Royal Society B*, **275 (1643)**, 1587–1593.

Sutton, M.D., Briggs, D.E.G, Siveter, David J., et al. (2001a) Methodologies for the visualization and reconstruction of three-dimensional fossils from the Silurian Herefordshire Lagerstätte. *Palaeontologia Electronica*, **4 (1)**, 1A.

Sutton M.D., Briggs, D.E.G., Siveter, David J., et al. (2001b) An exceptionally preserved vermiform mollusc from the Silurian of England. *Nature*, **410 (6827)**, 461–463.

Sutton M.D., Briggs, D.E.G., Siveter, David J., et al. (2001c) A three-dimensionally preserved fossil polychaete worm from the Silurian of Herefordshire, England. *Proceedings of the Royal Society, B*, **268 (1483)**, 2355–2363.

Sutton, M.D., Garwood, R.J., Siveter, David J., et al. (2012a) SPIERS and VAXML; a software toolkit for tomographic visualisation and a format for virtual specimen interchange. *Palaeontologia Electronica*, **15 (2)**, 5T.

Sutton, M.D., Briggs, D.E.G., Siveter, David J., et al. (2012b) A Silurian armoured aplacophoran and implications for molluscan phylogeny. *Nature*, **490 (7418)**, 94–97.

Swanson, L.W. (1983) Liquid metal ion sources: mechanism and applications. *Nuclear Instruments and Methods in Physics Research* **218 (1–3)**, 347–353.

Uchic, M.D., Holzer, L., Inkson, B.J., et al. (2007) Three-dimensional microstructural characterization using focused ion beam tomography. *MRS Bulletin*, **32 (5)**, 408–416.

Volkert, C.A. & Minor, A.M. (2007) Focused ion beam microscopy and micromachining. *MRS Bulletin*, **32 (5)**, 389–399.

Wacey, D., Kilburn, M.R., Saunders, M., et al. (2011) Microfossils of sulphur-metabolizing cells in 3.4-billion-year-old rocks of Western Australia. *Nature Geoscience*, **4 (10)**, 698–702.

Wacey, D., Menon, S., Green, L., et al. (2012) Taphonomy of very ancient microfossils from the ~3400 Ma Strelley Pool formation and ~1900 Ma Gunflint formation: new insights using a focused ion beam. *Precambrian Research*, **220–221**, 234–250.

Wirth, R. (2004) Focused ion beam (FIB): a novel technology for advanced application of micro- and nanoanalysis in geosciences and applied mineralogy. *European Journal of Mineralogy*, **16 (6)**, 863–876.

Further Reading/Resources

Briggs, D.E.G., Siveter, David J., Siveter, Derek J., et al. (2008) Virtual fossils from a 425 million-year-old volcanic ash. *American Scientist*, **96 (6)**, 474–481.

Giannuzzi, L.A. & Stevie, F.A. (eds) (2005) *Introduction to Focused Ion Beams: Instrumentation, Theory, Techniques and Practice*. Springer, New York.

Herbert, M.J. (1999) Computer-based serial section reconstruction. In: Harper, D.A.T. (ed), *Numerical Palaeobiology: Computer-Based Modelling and Analysis of Fossils and their Distributions*, pp. 93–126. Wiley, Chichester.

Kummel, B. & Raup, D. (eds) (1965) *Handbook of Paleontological Techniques*. Freeman, San Francisco.

Orloff, J., Utlaut, L. & Swanson, M.W. (2003) *High Resolution Focused Ion Beams: FIB and its Applications*. Kluwer Academic/Plenum Publishers, New York.

Yao, N. (ed) (2007) *Focused Ion Beam Systems: Basics and Applications*. Cambridge University Press, Cambridge.

3 Non-Destructive Tomography

Abstract: Non-destructive tomography is a group of scanning technologies which can create tomographic datasets of a fossil without causing any damage. This varied collection of approaches has a diverse range of underlying principles. X-ray computed tomography builds tomograms from radiographs of a rotating sample by mapping its X-ray attenuation in three dimension. This versatile tool, introduced in depth herein, is the most widespread technique for the tomographic study of fossils. Neutron tomography is similar to CT, but free neutrons provide the penetrating radiation. While challenging and rare in palaeontology, it could be a valuable option for some large fossils. Magnetic resonance imaging uses strong magnetic fields which can map the nuclei of light elements such as hydrogen within a sample and is thus better suited to imaging biological soft tissues than fossils. Finally, optical tomography techniques use visible light to acquire tomograms, but require a translucent matrix or macerated fossils.

3.1 Introduction

Non-destructive tomography comprises a collection of scanning methods which can be used – through the interaction of electromagnetic radiation or subatomic particles with matter – to create tomographic datasets without harming a sample. This is particularly valuable in palaeontology where fossils, especially those preserved in three dimensions, are seldom commonplace. For this reason, destructive tomography is a last resort; one, or several, of the techniques outlined in this chapter will be applicable to the majority of fossils. The techniques are very varied in underlying theory, practice, accessibility, time requirements and, indeed, merit for different fossils. Much of the chapter is dedicated to CT scanning (Section 3.2) which is the most widespread tomographic technique for the study of fossils. In

Techniques for Virtual Palaeontology, First Edition. Mark D. Sutton,
Imran A. Rahman and Russell J. Garwood.
© 2014 John Wiley & Sons, Ltd. Published 2014 by John Wiley & Sons, Ltd.

addition to history (Section 3.2.1) and background principles (Section 3.2.2), a range of different forms of CT are introduced in varying levels of detail (Sections 3.2.3–3.2.7), followed by a discussion of reconstruction and associated considerations (Sections 3.2.8–3.2.11), likely future directions (Section 3.2.12) and three case studies (Section 3.2.13). This is followed by sections on neutron tomography (NT) (Section 3.3), magnetic resonance imaging (MRI) (Section 3.4) and optical tomography (Section 3.5), all of which are less commonplace, but can be of great utility for fossils with specific forms of preservation.

3.2 X-Ray Computed Tomography

3.2.1 Introduction to CT

X-ray computed axial tomography, commonly CT, is a scanning (i.e. non-destructive) technology that utilizes X-rays to create a tomographic dataset. A number of different forms of CT scanning exist that differ in their resolution and applications. Some are laboratory scale (medical, micro- and nanotomography), while others employ substantial infrastructure (i.e. a **synchrotron**). All variants share common principles: an X-ray source and detector are used to acquire radiographs (or projections) of an object at multiple angles. From this, a sequence of parallel and evenly spaced tomograms is created, which maps the X-ray attenuation within a sample. X-ray attenuation is a function of numerous **material properties**, principally elemental composition/concentration and density. In fossils, minerals can often be differentiated allowing interior and exterior structures digitally visualized in three dimensions (see Chapter 5).

Fossils are frequently difficult to scan, typically being dense, highly variable in shape and composition and often lacking clear or consistent contrast between specimen and matrix. Thus, while CT is a very powerful technique for the study of fossils, its full potential can only be realized with the aid of a comprehensive understanding of how to obtain the best scans for a given sample. This section is intended to provide such an understanding, and it is worthy of note that even the best scanning parameters for any given fossil can provide challenging data; visualization of such fossil CT datasets may require considerable **virtual preparation** work (see Section 5.3.4.3).

Section 3.2.2 outlines a brief history of the technique, and Section 3.2.3 gives a grounding in the physics of the interaction of X-rays with matter, details of which underpin the selection of experimental parameters in any given CT scan. Section 3.2.3 discusses X-ray microtomography, the workhorse, and most widely available form of CT for fossil studies. This is followed by overviews of medical CT (Section 3.2.5), **nanotomography** (Section 3.2.6) and synchrotron-based CT (Section 3.2.7), slice reconstruction from projections (Section 3.2.8) and associated artefacts (Section 3.2.9), phase-contrast tomography (Section 3.2.9), and then an overview of scanning considerations (Section 3.2.10) and possible future directions for X-ray-based tomographic techniques (Section 3.2.11).

3.2.2   History

CT was originally developed in the 1970s as a clinical diagnostic tool in which X-ray source/detector pairs rotate around a patient (Hounsfield 1973). This innovation was made possible by increasing computing power and improvements in X-ray imaging and resulted in the award of a Nobel Prize in medicine to Godfrey Hounsfield and Allan Cormack in 1979. Full accounts of its development can be found in Friedland and Thurber (1996) and Petrik et al. (2006); Webb (1990) provides a comprehensive overview of its early history. Shortly afterwards, the applicability to other problems became apparent, and the technique was co-opted to great success in a wide range of different fields where non-destructive three-dimensional (3D) reconstruction is required.

3.2.2.1   CT in the Earth Sciences

Until the last few years of the 20th century, CT in the earth sciences was an experimental technique. Early work used sequential medical CT scanners – those in which each 360° rotation was followed by a brake and then reversal of the rotating components (the *gantry*). A new slice was created by moving the scanned object along the longitudinal axis during the brake. Furthermore, to minimize patient exposure to ionizing radiation, such systems employed a limited dose of low-energy X-rays, restricting sample sizes for geological applications (Arnold et al. 1983). Nevertheless, the potential of CT to earth science topics was clear from early studies, which included the analysis of meteorite inclusions (Arnold et al. 1983), reconstruction of lithological cores for petrophysics/petroleum (Wellington and Vinegar 1987), application in sedimentology (Renter 1989) and soil bulk density analysis (Petrovic et al. 1982). Early palaeontological work with CT involved large vertebrate fossils, which are both a suitable scale for medical CT scanners and the smallest conceptual leap from CT's diagnostic origins. Palaeoanthropology was hence among the first adopters of CT technology (Wind 1984; Zonneveld and Wind 1985; Zonneveld et al. 1989; Wind and Zonneveld 1989). Other early studies included the qualitative and quantitative comparison of fossil and extant bone (Tate and Cann 1982), the investigation of a matrix-filled Miocene ungulate skull (Conroy and Vannier 1984) and the analysis of *Archaeopteryx* (Haubitz et al. 1988). This early CT development was accompanied by advances in visualization techniques (see Chapter 5). The next 20 years saw CT evolve from an experimental and exploratory technique to a well-established one; its application is now commonplace, and its utility for the study of three-dimensional fossils widely recognized. This period has seen a division into three primary forms of data acquisition: medical, micro- and synchrotron CT.

3.2.2.2 Medical Scanners

Although representing a small proportion of palaeontological studies conducted today, factors including familiarity and accessibility account for the continued use of medical CT scanners in analyses of fossils. There have been significant advances in medical scanner technology since the late 1980s, which have considerably improved their performance, most notably resolution and acquisition time. The development of slip rings – electrical connections via stationary brushes in contact with rotating circular conductors – removed rotation limitations caused by cabling and facilitated continuous rotation of the moving components. This contributed to the development of spiral (or helical) scanning, in which the specimen is moved smoothly in the z-direction within a continuously rotating gantry, tracing a spiral around the scanned object (Kalender 2006). This approach treats data as a three-dimensional **volume**, from which specialized reconstruction algorithms employing interpolation are able to recover slice data (Fuchs et al. 2003). More recently, the addition of multi-row detector arrays which allow numerous slices to be collected each rotation has decreased scan times further still (Schaller et al. 2000). Modern studies which employ medical scanners will hence usually be utilizing **spiral CT** scanning, although few details beyond the model or location of the scanner are generally given. Spiral CT is most commonly reported in CT work on vertebrates due to the scale of the fossils (e.g. Rogers 1998; Sanders and Smith 2005; Marino et al. 2003), but invertebrate examples also exist (e.g. Beuck et al. 2008).

3.2.2.3 Microtomography/Micro-CT

By the 1980s, it had become clear that the limitations on X-ray energy, maximum dosage and imaging time required for medical scanners restricted the applications and resolution of the technique. Accordingly, there was a rapid move towards the development of non-medical systems for research and industrial scanning. Of the associated technologies, X-ray microtomography (μCT/XMT), in which radiographs of a rotating sample are collected on a two-dimensional detector array, has had the greatest impact. It has allowed smaller samples such as invertebrates to be studied at high resolutions (i.e. <100 μm). Elliott and Dover (1982) demonstrated the technique by modifying a scanning X-ray microradiography system (reported in Elliott et al. 1981). They successfully created a cross-sectional image of the aragonite shell of the extant snail *Biomphalaria glabrata*, with a **voxel** size of 12 μm. The majority of early studies were conducted with custom-built scanners (e.g. Feldkamp et al. 1989). Relatively few examples of high-resolution computed tomography exist in the geosciences until the early 2000s (although see Rowe 1996; Cifelli et al. 1996). During this time, XMT tomographs became commercially available for the first time – early studies used these scanners to study bone structure (Rüegsegger et al. 1996). By the turn of the millennium, micro-CT (or the related high-resolution industrial microtomography) was an established technique which was beginning to be applied in the geosciences (Kuebler et al. 1999; Proussevitch et al. 1998; Denison et al. 1997). Trailblazers in the development of CT as a geosciences tool were the University of Texas High-Resolution X-ray Computed Tomography Facility (Ketcham and Carlson 2001; Carlson et al. 2003), who were also responsible for much of the early work in palaeontology (Rowe et al. 2001; Maisey 2001; Dominguez et al. 2002), although examples from elsewhere do exist (e.g. Thompson and Illerhaus 1998). Many of these examples involved vertebrate material, but the CT revolution has seen this scope widen considerably to a wide range of invertebrates including, to list a few, echinoderms (Rahman and Zamora 2009), molluscs (Vendrasco et al. 2004), corals (Molineux et al. 2007) and arthropods (Garwood and Sutton 2010), as well as plants (DeVore and Kenrick 2006) and microfossils (Görög et al. 2012). The last decade has seen micro-CT develop rapidly: scanners have become more available and widespread (Abel et al. 2012), and the technique relatively cheap, with scans typically costing less than £100. The technique has been continuously refined, with technological advances improving the resolution of lab-based systems to sub-micrometre levels (Dunlop et al. 2012), and allowing the penetration of denser specimens (e.g. metal-rich nodules; Garwood et al. 2009).

3.2.2.4 Synchrotron CT

A synchrotron is a cyclic particle accelerator in which a beam of high-energy charged particles, normally electrons, are kept moving in a circular path using electromagnetic fields. The concept was independently proposed by Veksler (1944) and McMillan (1945), and the first synchrotron was constructed by General Electric in the USA shortly afterwards. When the direction of the beam is changed (i.e. the particles are accelerated), X-rays are emitted with a wide range of energies and high photon flux. Elder et al. (1947) announced the discovery of synchrotron radiation which was, ironically, viewed at first as an unavoidable form of energy loss during electron acceleration for collision with a **target metal** to create intense X-rays. Nevertheless, their utility was soon recognized, and numerous synchrotron light sources were constructed over the next four decades, which are now employed for a huge range of X-ray techniques. Thus, the infrastructure was in place shortly after the development of XMT to use synchrotron radiation in tomographic studies (Flannery and Deckman 1987). By the mid-1990s, an initial reliance on X-ray attenuation was augmented by the development of phase-contrast imaging (Snigirev and Snigireva 1995; Cloetens et al. 1996), facilitating studies of samples with low **attenuation** contrast. The last decade has seen the development of an increasing number of third-generation synchrotrons, conceived to produce bright X-rays, and allowing a surge of high-profile studies conducted with synchrotron micro-CT in the last decade. Studied fossils include the Doushantuo fauna (Hagadorn et al. 2006) and other Neoproterozoic fossils (Donoghue et al. 2006), hominids (Chaimanee et al. 2003), ammonites (Kruta et al. 2011) and seeds (Friis et al. 2007).

3.2.3   X-Rays and Matter

All forms of X-ray computed tomography rely on the interaction of X-rays and matter to produce tomographic datasets of an object. The majority create attenuation maps, where attenuation is the loss of intensity in X-radiation as it passes through a medium, resulting from absorption and/or scattering. Numerous forms of CT exist, varying primarily in the means by which X-rays are generated and data is acquired. This section is intended as a practical introduction to CT, and therefore, the physics of X-rays and their interaction with matter are dealt with briefly – full treatments can be found elsewhere (Als-Nielsen and McMorrow 2011). However, an understanding of X-ray matter interaction underlies the choice of scanning technologies and parameters and is thus central to successful CT scanning.

3.2.3.1   X-Rays, Energy and Intensity

X-rays are an electromagnetic radiation with a wavelength between 0.01 and 10 nm (by contrast the wavelength of visible light is 380–740 nm). The energy of a photon is commonly measured in electron volts (eV). One eV is 1.602×10^{-19} J, which is the kinetic energy gained by a single electron accelerated across an electrical potential of 1 V. Photon energy is inversely proportional to wavelength (i.e. longer wavelengths correspond with lower-energy X-ray photons). The energy of X-rays determines their degree of attenuation in a material, the mechanism by which this occurs, and their penetrative ability – the latter is almost always a consideration with fossils, which are often both dense and large. Higher-energy X-rays are less sensitive to changes in composition and (to an extent) density, but penetrate objects more effectively than those of low energy. Accordingly, the X-ray spectrum is often bisected into waveforms with wavelengths between 0.1 nm (12.4 keV) and 10 nm (124 eV), called **soft X-rays**, and **hard X-rays**, with wavelengths of 0.01 nm (124 keV) to 0.1 nm. The former generally lack the penetration required for palaeontological tomography applications. The X-ray intensity (approximating to the number of X-ray photons per unit time) is also important in tomography; higher beam intensities result in better signal-to-noise ratios (SNR). However, high-intensity beams can require a source with a larger focal spot, which lowers (coarsens) resolution (Section 3.2.4.2).

3.2.3.2   Attenuation Mechanisms

X-rays can interact with matter in numerous ways. At the energies used for CT scanning, there are two principal attenuation mechanisms: the **photoelectric effect** and **Compton scattering** (Buzug 2008). The former occurs when an X-ray photon's energy slightly exceeds the binding energy of an atomic electron, which it then liberates as a **photoelectron**. The X-ray photon, having transferred its energy, ceases to exist, and the photoelectron's ejection creates a vacancy. With lower-energy X-rays, this occurs in outer shells but, with increasing photon energy, can instead occur with inner-shell

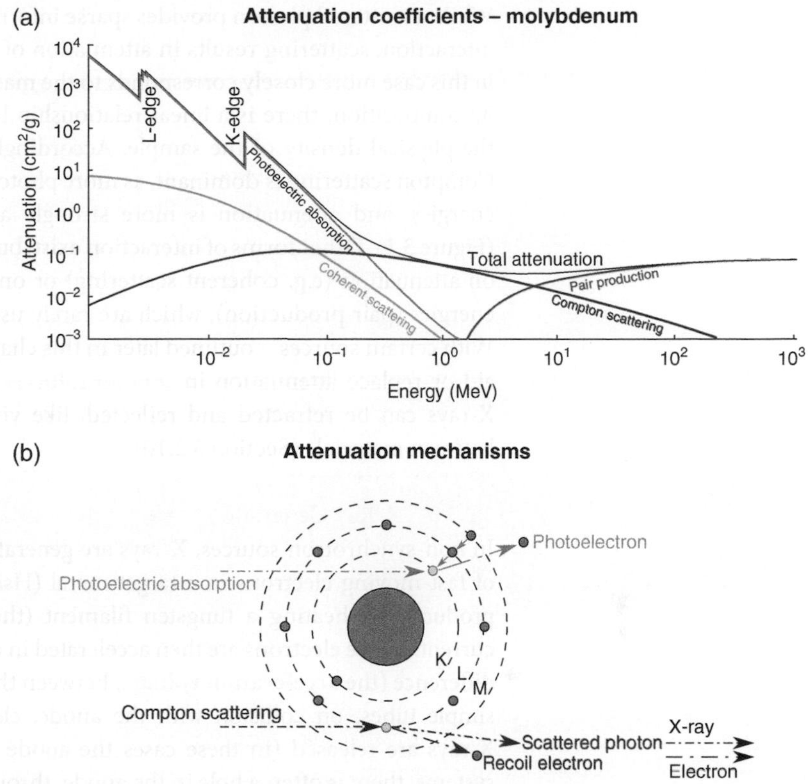

Figure 3.1 (a) The attenuation (y-axis, cm^2/g) of molybdenum at different energies (x-axis, MeV) with contributing mechanisms of attenuation. Note absorption edges and varying contributions of photoelectric (grey) and Compton effects (red). Pair production occurs at higher energies than those found in a CT lab. Graphed with data from the NIST XCOM database (www.nist.gov/pml/data/xcom). (b) The two attenuation mechanisms dominant in typical CT scans.

electrons: if the X-ray energy is greater than the binding energy of an inner electron (K-shell), as is often the case for hard X-rays, the inner shell is preferentially affected by this process. When X-ray energy reaches the binding energy of any given shell – which is element dependent – there is a distinct jump in the **attenuation coefficient** of that element. This is known as an absorption edge (**K-edge** for K-shell, L-edge for L-shell, etc.). Electrons from outer orbitals will fill any vacancy. The vacancy consequently cascades to an outer shell to be filled with a free environmental electron. However, the likelihood of photoelectric absorption decreases with increasing excess photon energy, and accordingly, attenuation decreases with increasing energy in any given element, giving the attenuation coefficient a sawtooth appearance with increasing energy (Figure 3.1). This mechanism is dominant at lower energies (those close to the binding energies of inner-shell electrons, generally <50 keV for most palaeontological purposes), and the photoelectric effect increases proportionally to the fourth or fifth power of the atomic number of an element. Thus, at low energies, the attenuation map is strongly affected by the chemical composition of the scanned object.

Where photon energy greatly exceeds the binding energy of the electrons in an atom, Compton scattering dominates. This mechanism arises from photons and electrons colliding, resulting in a partial photon energy loss (then scattering or deflection) and freeing a 'recoil' or secondary electron.

While a scattered photon provides sparse information on the location of the interaction, scattering results in attenuation of the X-rays. The attenuation in this case more closely corresponds to the mass density of the sample than its composition: there is a linear relationship between the attenuation and the physical density of the sample. Accordingly, at higher X-ray energies, Compton scattering is dominant, as more photons exceed electrons' binding energies, and attenuation is more strongly affected by physical density (Figure 3.1). Other forms of interaction exist, but have relatively little impact on attenuation (e.g. coherent scattering) or only occur at very high X-ray energies (pair production), which are rarely used in tomographic imaging. With certain sources – outlined later in this chapter – other interactions can aid or replace attenuation in **tomographic reconstruction**. For example, X-rays can be refracted and reflected, like visible light, at the interfaces between materials (Section 3.2.10).

3.2.3.3　X-Ray Generation

In non-synchrotron sources, X-rays are generated through the deceleration of fast-moving electrons in a target metal (Hsieh 2003). The electrons are produced by heating a tungsten filament (the cathode) with an electric current. These electrons are then accelerated in a vacuum by a high potential difference (the **acceleration voltage**) between the filament and the anode. In simple tubes, on collision with the anode, electrons are decelerated and X-rays are released (in these cases the anode is the target). In micro-CT systems, there is often a hole in the anode, through which the electron beam passes, and after which it is directed by deflecting magnets and an electromagnetic lens. This focuses it onto the target metal, which produces X-rays (Section 3.2.4.2, Figure 3.4). In both instances, an X-ray photon's energy relies upon the incident electron's velocity, which is – in turn – determined by the acceleration voltage between the cathode and the anode. Hence, acceleration voltage dictates the maximum X-ray energy of a source – for example, if it is 125,000 V (125 kV), the most energetic X-rays will be 125 keV (Figure 3.2). However, the output of a typical lab source is a broad energy spectrum, often described in terms of its highest energy (keV or MeV), but with a maximum intensity often found at energies significantly below this.

　This spectrum comprises two components. The first is the **bremsstrahlung** ('braking radiation' in German) – a continuous curve between the minimum and maximum X-ray energies. This is generated as a result of the deceleration of electrons by the electric field of nuclei in the metallic target. Lost energy is released as X-ray wavelength radiation; the greater the electron's energy loss, the higher the energy of the resulting X-ray photon. The upper limit of the spectrum is caused by the direct collision of an electron with the nucleus; here the entirety of the kinetic energy is converted to an X-ray photon. The bremsstrahlung is punctuated by **characteristic radiation**: energy peaks which are unique to any given element and conceptually similar to the absorption edges of attenuation profiles. Characteristic radiation results from high-speed incident electrons interacting with atomic electrons in the target, ejecting some from their shell. Characteristic X-rays (Figure 3.2) are emitted

X-ray generation – tungsten

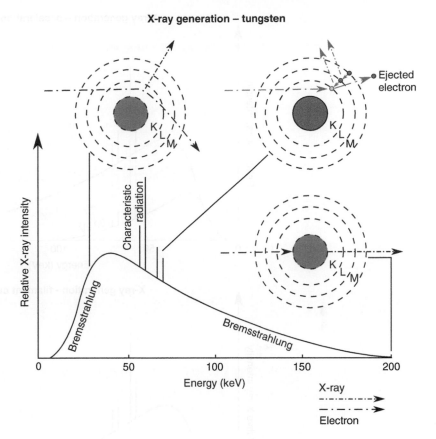

Figure 3.2 Electron–target interaction within an X-ray source, and the resulting energy spectrum, based on a tungsten target metal. Bremsstrahlung (red) and characteristic (black) radiation are present in lab sources of this form – they are formed by differing interactions with high-speed incident electrons and the target metal's atoms. Interactions shown represent low and maximum energy bremsstrahlung and characteristic peaks.

when outer-shell electrons fill this vacancy, the strongest being typically the **K-alpha doublet**, which represents electron transitions from a p orbital of the L-shell to the vacated K-shell, with a weaker K-beta (M to K transition).

In addition to dictating the maximum energy of the bremsstrahlung X-rays, changing the acceleration voltage will also modify the shape of the X-ray spectrum: raising it will increase the average X-ray energy (i.e. the bremsstrahlung intensity peak will move right; Figure 3.3). Increased acceleration voltage will also increase the intensity of an X-ray source and thus the amplitude of its spectrum. In addition, **filament current** impacts on intensity – a greater current increases X-ray emission at all energies. The energy of the bremsstrahlung peak intensity and characteristic radiation will not change, but the spectrum's amplitude will increase (Figure 3.3). The majority of interactions in the target do not produce X-rays – instead, they produce heat. Accordingly, the limit of the X-ray intensity is dictated by the maximum power the target can withstand without overheating. Most X-ray sources have cooling systems (usually water) to counteract the heating, and some dissipate the heat over a larger volume by rotating the target (often also the anode, in these systems).

Numerous metals can be used in the target: a metal with a higher atomic number increases the energy of both the peak intensity of the

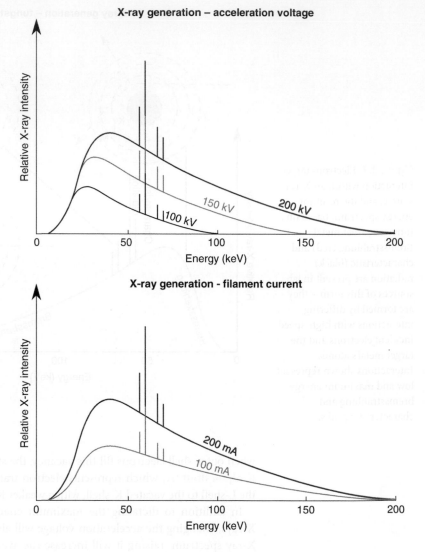

Figure 3.3 Energy spectra from a lab source under different acceleration voltages and filament currents.

bremsstrahlung and characteristic radiation. Among the most common are, in order of decreasing wavelength of K-alpha characteristic radiation and thus increasing energy, chromium (~5.4 keV), iron (~6.4 keV), cobalt (~6.9 keV), copper (~8.0 keV), molybdenum (~17.5 keV), silver (~22.2 keV) and tungsten (~59.3 keV). As atomic number increases (to the right of this list), so does the intensity of the source. Many scanners – especially microtomography machines, often designed for versatility – provide multiple target metals.

3.2.3.4 X-Ray Detection

To capture an X-ray projection, a detector is needed. X-rays are not measured directly, but are rather detected through their interactions with other materials – for example, by the emission of photoelectrons or visible light. Detectors can influence the quality of a scan through both their size and their efficiency in detecting the energy spectrum generated by the source. Early scanners used gas-based systems (using the principle of a Geiger–Müller tube), and some medical scanners still use high-pressure xenon detectors. These are relatively rare, however, and the vast majority of scanners, especially micro-CT and synchrotron systems, rely on solid-state, *scintillator* detectors. These comprise a luminescent scintillator medium, such as caesium iodide (CsI) or calcium tungstate ($CaWO_4$), which re-emit absorbed X-rays as light. Multiple scintillator materials are available, with varied efficiencies and response/decay times: these are built into scanning systems and are thus a parameter which palaeontologists generally need not consider. The scintillator can be followed by an optical relay element to focus and sometimes amplify the signal. All systems then possess photodetectors such as photodiodes, photomultipliers or charge-coupled device (CCD) panels (much like a digital camera) to convert the light signal into an electrical signal for further processing.

3.2.4   X-Ray Microtomography

Micro-CT possesses a multitude of possible applications, necessitating a large range of scales, and is generally divorced of limitations such as those imposed in medical CT by radiation dosage. Accordingly, scanners come in a large variety of different forms. They typically have minimum voxel sizes between 1 and 100 μm, although these are not well-defined limits – gradations exist between micro- and nano-CT at one extreme and medical/industrial CT on the other. XMT tomographs employ *volume scanning*, in which a sample is rotated between an X-ray source and a two-dimensional detector panel array. They offer a larger variety of scanning parameters than most other forms of CT, including X-ray source target material, source voltage and current, exposure time, frame averaging and number of projections. All are discussed in the current section, along with a practical guide to conducting a scan, intended to be applicable to as many XMT set-ups as possible. Authors – especially those unfamiliar with the CT – often provide methods sections that lack vital scanning parameters. This can make the work unrepeatable and omits information that could provide valuable guidance for other workers; we hope this introduction to scanning will provide the information required for authors to forestall this issue.

3.2.4.1   Resolution

Source spot size, specimen size and detector dimensions are the main controls on resolution in any given scan. In much palaeontological scanning, resultant tomographic datasets will be isotropic – that is, the *xy*- and *z*-resolutions are the same, and the resulting data can be reconstructed with cubic voxels (see Section 5.3.4.1). Accordingly, the detail in a scan is normally reported with the voxel size (i.e. the distance between each voxel centre, rather than the maximum dimension of each voxel). This is frequently provided in files accompanying the projections of a scan, which record scanning parameters. Each line of pixels on the detector panel will be reconstructed as a single tomogram in a tomographic dataset. Therefore, a 2000 × 2000 pixel detector panel would produce a total of 2000 slices, with each tomogram consisting of 2000 × 2000 pixels (i.e. a 2000 × 2000 × 2000 voxel dataset). The use of a cone beam in non-synchrotron sources provides simple geometric magnification of the object: the closer it is to the target, the greater its magnification on the detector panel (Davis and Wong 1996), and hence the smaller the voxel size. However, voxel size is also limited by the size of an object – all parts of the sample should remain in the field of view of the detector through a 360° rotation; violations can cause detrimental artefacts. The voxel size is hence coupled to the size of the object scanned: in a 2000 × 2000 pixel detector panel, the smallest voxel size possible would be 1/2000th of the maximum dimension of the object. For many fossils, this stricture provides practical limits for the detail visible in a scan.

Spatial resolution is the ability to resolve small, closely positioned objects as separate forms. It depends on many factors including properties of the source, sample, detector and reconstruction algorithms. The spatial resolution of a scan is sometimes provided as a multiple of the voxel size – for example, to resolve two objects as distinct forms may require five voxels. However, a more accurate estimate can be obtained using a phantom – a specially designed object with patterns of high and low attenuation bars at different scales (Hsieh 2003). This can be scanned under identical settings to the specimen, and from the smallest-resolvable patterns, the true spatial resolution can be identified. This is usually impractical for palaeontological scans, and because voxel size is normally provided in the scan information, it is generally preferable. Note, however, that there is an absolute limit to the spatial resolution of a scan – the spot size of the X-ray source.

3.2.4.2 Source

If very small objects are placed close enough to the source that the voxel size is smaller than that of the focal spot on the target, further magnification does nothing to improve the spatial resolution or clarity of the resulting dataset. The focal spot size determines the possible number of source to detector paths at any given point within a scanned object, and further magnification will lead to blurred edges (penumbra effects; Withers 2007). Spot size – typically between 1 and 4 μm – is only a real consideration for very high-resolution scans (especially those in nano-CT; see Section 3.2.6). Numerous factors can affect the spot size of a source (and thus the absolute resolution of a system). Source parameters can also dictate the maximum energy/intensity of X-rays (see Section 3.2.3.3) and hence their penetration depth and accordingly a scanner's maximum sample size. X-rays are emitted by a region in the target known as the X-ray excitation volume. A low acceleration voltage minimizes this volume (but will result in low intensity), as does using target metals of higher atomic number. Two forms of source also exist (Schambach et al. 2010). In a transmission source, the target (usually

X-ray generation – transmission target

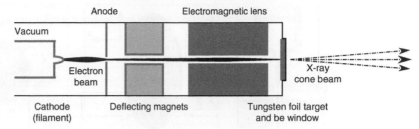

X-ray generation – reflection target

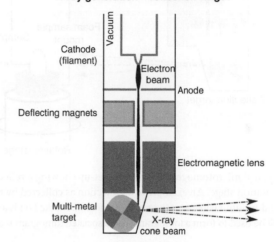

Figure 3.4 A transmission and reflection microfocus target of the type, and in the orientation, usually found in XMT machines.

tungsten) is a thin sheet of metal (Figure 3.4). The electrons are accelerated parallel to the orientation of the cone beam and create a small X-ray excitation volume equal to the thickness of the sheet. The resulting spot size is small, but emits lower-intensity X-rays. In micro-CT, transmission targets are generally used for objects where spot size is the limiting factor in the resolution. As such specimens are by definition small, the low intensity is seldom problematic. In contrast, a reflection target is positioned at an angle to the electron beam (Figure 3.4), creating a greater X-ray excitation volume and hence higher-intensity X-rays, which originate from the surface that the beam impinges upon. This higher intensity renders refection targets more suitable for fossils, but the spot size is larger. In micro-CT scanners, the reflection source often has strips of different metal targets that can be rotated by hand, allowing for easy alteration of the target material.

3.2.4.3 Sample Preparation

There is a common workflow to setting up a CT scan on any given system, the first element being sample preparation. In the vast majority of cases, this will be solely a matter of mounting the specimen. We recommend orienting the sample's longest axis vertically, as this reduces the maximum horizontal thickness which the X-ray beam will be required to penetrate (Figure 3.5),

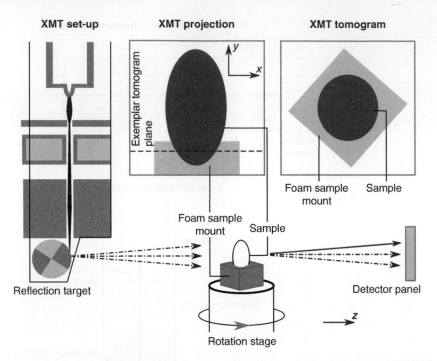

Figure 3.5 A typical microtomography scanning set-up showing a reflection target, and a florist's foam mounted sample on the rotation stage. An example of a projection as collected by the detector panel is also shown, with the specimen and mount visible – note sample orientation. Marked on is a single line of pixels which – with the aid of all projections – will be used to form a tomogram. The associated tomogram is also shown.

reducing anisotropy and associated artefacts and increasing SNR. Having any part of the stage in the field of view can also introduce artefacts and should be avoided. It is possible to scan objects with host rock (or any other elements which are not of interest) out of the field of view, but depending on the reconstruction algorithm, this can introduce artefacts that appear as a bright halo around the edges of the scan. This can be countered with the aid of simple modifications to the data prior to reconstruction (Section 3.2.8) in some homogeneous samples (Dierick et al. 2007). Otherwise, however, this form of region of interest scanning or *local tomography* is currently experimental and relies on custom-built reconstruction algorithms (Rashid-Farrokhi et al. 1997; Wang et al. 1999), making it hard to access for palaeontological samples.

It is essential that specimens do not move, settle or wobble during scanning; even a small shift can ruin the data and necessitate a re-scan. Large (>0.15 m) specimens can simply be placed in a bucket or jug directly on the rotation stage to keep them in the correct orientation. Scanners will usually have an upper sample size limit based on detector dimensions or cabinet size, and many stages have a 10–20 kg weight limit. Both should be checked prior to placing a large fossil on the manipulation arm. For smaller specimens, mounting is usually achieved with a soft material of low X-ray density, such as polystyrene or florist's foam; these can easily be cut to shape to house a specimen. Where sample size is the limiting factor on resolution (see Section 3.2.4.1), removing excess host rock around small samples will be beneficial, although there are often practical reasons not to perform it (e.g. the danger of damaging the specimen itself, or curatorial policy). Trimming is especially valuable in highly anisotropic (slablike) samples where one axis is much shorter than the other two, such as fossils preserved in shale. Here the variation in penetration between the longer axes (typically parallel to the bedding plane) and the short axes can reduce reconstruction quality; cutting the specimen so only one axis remains relatively long provides a solution, as the long axis can be oriented vertically. Alternative approach in such cases (in non-phase-contrast analyses) is to reduce variation in penetration by scanning the sample buried in another substance, creating a more isotropic sample with less angular disparity in attenuation values. For low-density samples, this may be flour, while for those of higher density, it could be sand. This is also beneficial in dense samples where the air–sample interface can be obscured by streak artefacts (Section 3.2.9) due to the sharp change in attenuation. The smallest samples will often need to be held in place with florist's foam or cotton wool within another holder, such as a measuring cylinder, the tip of a pipette or adhered to the head of a needle. For attaching samples, water-soluble glue or double-sided tape is preferable to Blu-tack, which is very X-ray dense, and can hence introduce artefacts. Many custom-built sample holders, including clamps, dowels and wedges, also exist.

Note: Micro-CT is non-destructive, but we are aware of at least one case where it can be detrimental to future analyses. Electron spin resonance (ESR) dating of fossils, typically teeth (Grün et al. 1987; Grün and Stringer 2007), relies upon radiation exposure to estimate a sample's age. As X-rays are a form of ionizing radiation, a CT scan will render subsequent ESR dating incorrect.

3.2.4.4 Positioning the Sample

Mounted specimens are placed on the rotation stage, whose centre of rotation (COR) is aligned with the centre of the detector panel. When correctly positioned, the COR of the specimen should be immediately above the stage COR: positioned in this way projected images of the specimen rotate around the centre of the detector panel through a 360° rotation. An incorrectly centred specimen will move from laterally during rotation.

To correctly position the specimen, ensure X-rays are off using the scanner software, and open the door (see also for safety information Section 3.2.11.2). The sample should be placed on the stage, door closed, X-rays turned on, and a live projection image inspected. The manipulator position can then be adjusted, either via joystick or software – while the manipulator arm is in motion, it is advisable to carefully watch through the viewing window in the door in addition to checking the projection, as collision of the arm with the source or detector is distinctly undesirable. To correctly position the specimen in the vertical direction (i.e. by adjusting the y-axis), its middle should be aligned with the middle of the detector panel (i.e. so half is above the detector midline, and half below). This will often highlight an incorrect z-position (or zoom). If zoom is too high, the sample projection will be larger than the panel and hence be cut off at the top and bottom; if too low, there will be wide top and bottom zones with no sample. Zoom should be corrected by moving the arm backward or forward in the z-direction.

The sample is then centred in the x-axis (i.e. across the detector) for 360° of rotation by moving it *relative to – until aligned with – the COR* of the stage. Some but not all systems aid this process by mounting samples on a centred structure such as a thin metal rod. In some scanners, there is a second sample stage above the rotation stage facilitating sample movement relative to the rotation centre. Where this is absent, position is altered manually by turning off X-rays and moving the sample. Once the sample's long axis (which should be vertical) is aligned with the centre of the detector panel at one angle, the process should be repeated at 90° to this. A sample aligned at orthogonal angles is centred; to confirm this, the sample should be rotated 360° to ensure it remains in the field of view throughout. Once the sample is correctly positioned, other scanning parameters can be set.

Note: Long specimens can be scanned in sections (allowing zoom levels based on the medium axis), and the resulting stacks concatenated. In the majority of systems, the detector is fixed in position, but in some it can be moved. This is equivalent to moving the sample in the z-direction: a more distant detector panel will allow for greater geometric magnification, but will require more energetic X-rays and could introduce more noise. Some systems allow lateral concatenation of scans; however, this is rare and prone to artefacts.

3.2.4.5 Scanning Parameters: Source

There are a range of scanning parameters that can be defined when setting up a CT scan – many combinations of these are likely to be equally effective for any given sample. The majority of scanners display a live histogram of the projection being collected by the detector panel and CCD, allowing the user to achieve the correct settings – conditions for scanning will be ideal when the histogram of the projection spans the majority of the detector's grey levels without hitting the extremes. A histogram with many peaks can also be considered a positive sign, suggesting a range of distinct attenuation materials in a sample. The best settings are generally the lowest-energy X-rays possible to penetrate a sample and create a reconstruction without **beam hardening** (Section 3.2.9.5) allowing for better elemental discrimination. Energy can be adjusted most easily using acceleration voltage and filament current (see Table 3.1 for a summary). Increasing voltage will increase the energy and intensity of the X-rays and thus usually increase the brightness of an image on the detector panel: to a rough approximation, this moves the entire histogram to the right without changing its shape. Many fossils will be scanned at acceleration voltages between 100 and 225 kV (the maximum for common scanner models). Increasing the filament current

Table 3.1 The impact of different parameters on X-ray energy and the projection histogram.

Factor	Effect on energy	Impact on projection histogram
Current	Increase in filament current leads to increased X-ray intensity.	Increase moves the right hand side to the right.
Voltage	Increase in acceleration voltage leads to increased X-ray energy and intensity.	Increase moves entire histogram to right.
Target metal	Energy and intensity increases with atomic number.	Increase in atomic number moves histogram to right.
Exposure time	None.	Increase moves entire histogram right.
Filtration	Reduces intensity (most markedly of low-energy X-rays).	Moves histogram left. Coupled with increased exposure, moves left of histogram right.

increases X-ray intensity and tends to extend the projection's histogram – the maximum and peak levels move to the right, but the minimum (left) remains in the same place. A typical current for scanning fossils is between 75 and 200 µA. The choice of target has the largest impact on scans through the characteristic X-rays of the metal – ideally these should be of an energy high enough to penetrate the fossil, but low enough to gain maximal compositional contrast. The choice of the target metal also has a small impact on spot size; however, apart from scans at the highest possible resolution, this is rarely an issue (Section 3.2.4.2). Our experience is that tungsten is the most suitable metal for almost all fossils, although molybdenum is preferable for amber.

An additional parameter is exposure time; the length of time over which each projection is captured. High exposure times result in bright images, but may saturate the detector panel (i.e. raise brightness above its maximum measurable threshold). If the exposure time is too low, projections will be too dark, contrast will be low, and noise in the reconstructed data will be high. With very X-ray dense fossils (such as iron-rich nodules), increasing the exposure time from a typical 275–500 to 1000–1500 ms, coupled with increasing the filter thickness (Section 3.2.4.6), can be the only means by which scanning is possible. Increasing exposure time does slow scan acquisition; to a first approximation total scan time is directly proportional to both number of projections and exposure time, although data-transfer bottlenecks and overhead time for performing a rotation step can complicate this relationship. Increased exposure does, however, improve the SNR as noise tends to cancel out over time. If noise precludes thresholding data (Section 5.3.4.2), lowering X-ray energy (to improve compositional contrast) and/or increasing exposure (to lower noise) can counteract this. If the lower energy is impossible due to beam hardening or lack of penetration, SNR can still be improved through frame averaging. Here, numerous frames (common options being 2, 4 and 8) are acquired at any given angle. These are then averaged, increasing and effective exposure time, and reducing noise (but again increasing scan time).

Another variable which affects the scan quality is the number of projections. Scan time is directly proportional to this (see preceding text), but the more projections, the better the SNR. More projections also enable reconstruction algorithms to more precisely differentiate fine details, although a law of diminishing returns applies here. The majority of systems will have a suggested minimum projection number based upon the detector size (usually to allow one projection per voxel at the edge of the data volume). If rapid scanning time is important (in palaeontology most frequently due to funding constraints), this can also be reduced – this risks increased noise and associated artefacts that appear first on the edges of the reconstruction and on sharp boundaries. Such artefacts can be minimized with novel, but slower and experimental, reconstruction algorithms (Section 3.2.8.2). The projections will by default normally occur over a 360° rotation. However, this is not a strict necessity – in lab-based XMT, the minimum required rotation is 180° plus the cone angle of the source. Lower angles are only recommended if the absolute minimum number of projections is required, for example, due to sample movement.

Some detectors offer a gain setting – using this amplifies the signal (and thus contrast), but also the noise of a projection. If very clean scans are required, setting this gain to low is advisable. High is better for high-contrast samples, as it will allow shorter exposures and quicker scans.

3.2.4.6 Scanning Parameters: Filters

Fossils are often very X-ray dense. As a result, less energetic X-rays may be incapable of penetrating them, and an artefact called beam hardening results (Section 3.2.9.5). Furthermore, the requirement that the entire sample always remains within the field of view necessitates regions where X-rays hit the detector panel without encountering the object (i.e. passing through the air space around the specimen; although see discussion regarding region of interest scanning in Section 3.2.4.3). When X-rays are powerful enough to penetrate a dense sample, those passing only through air can saturate the detector. This will result in poor details of the air–sample interface and can obscure internal anatomy that is not at the sample's centre. It also results in **calibration images** which are incapable of correcting pixel errors and thus prominent **ring artefacts** (Section 3.2.9). This can be overcome with the aid of filters: thin (0.1 to ~2.5 mm) metal films which can be placed directly in front of the source. Common metals are copper, tin, silver and brass, although composite filters which combining metals are also available. Filters increase the average energy of an X-ray source by removing lower-energy X-rays, but decrease the intensity (Figure 3.6). The higher-energy X-rays which remain are more penetrating, and thus – by increasing the current, voltage or

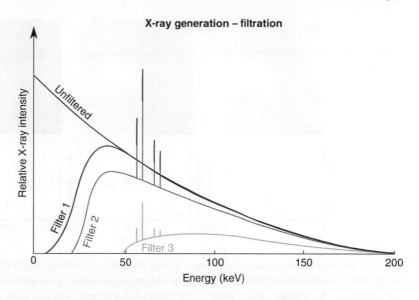

Figure 3.6 A diagram showing the effects of filtration on the energy spectrum of a lab source (based on a tungsten target metal). Spectra include an unfiltered beam (although note this will never occur, as even without a filter material air and the X-ray window will provide some filtration of a beam) and then filters of an increasing thickness from Filter 1 to Filter 3.

exposure – suitable scanning parameters can be found for the majority of fossils. Adding a filter will decrease the proportion of composition-based attenuation, however, as this is primarily associated with lower-energy X-rays, some of which would have penetrated the sample. The choice of filter material can facilitate a pseudo-monochromatic beam by choosing a metal with a K-edge at marginally higher energy than the target's K-alpha characteristic radiation (preferentially reduce K-beta and bremsstrahlung). However, this is usually unnecessary for conventional tomography; all that is required for an effective filter is a relatively low K-edge, allowing it to preferentially filter low-energy X-rays. A high K-edge filter will affect low energies less, and filtration will not have the desired effect. With this knowledge, trial, error and iterative improvement of filter thickness and material is usually both sufficient and effective. Copper, with strong absorption between 9 and 30 keV, is usually a suitable filter material for fossils.

3.2.4.7 Setting Scanning Parameters

Once a sample is prepared and positioned, acquiring a successful CT scan requires the user to alter the parameters described earlier – best judged from a live projection image – to match any given sample. Ideal scanning conditions can differentiate the mineral phases in a good proportion of fossils; they should provide the maximum possible contrast (widest spread of the image histogram), while still penetrating the sample, and not saturating the detector panel when the sample is not present. Selecting parameters can be challenging, and the help of an experienced tutor is advantageous. Table 3.1 recaps various scanning parameters.

Sensible histogram limits for a 16-bit (0–65,535) detector are 5000–60,000. To work out if a filter is required, it is best to first adjust voltage and current through the machine software without a filter. The effects of

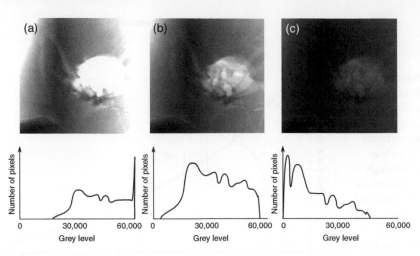

Figure 3.7 Projection images and histograms acquired with different settings. (a) With voltage, current or exposure too high, the lightest pixels are washed out, and the histogram possesses a peak around 60,000 (the maximum for many detectors). (b) Settings correct. Note the histogram falls between 8000 and 55,000 and shows a large number of peaks. (c) With voltage, current or exposure too low – the projection is too dark, and the histogram is bunched towards the left.

incorrect settings are shown in Figure 3.7. Experimentally varying these allows the user to seek suitable settings (using no filter maximizes composition-based attenuation so is preferable, if possible). During this process, sample should be rotated to its most X-ray dense orientation. If the darkest grey levels of the projection cannot be kept above 5000 without saturating the detector panel away from the sample, a thin filter should be added, and current, voltage and exposure adjusted accordingly (i.e. increased). This should be repeated with increasing filter thickness until darkest grey levels of over 5000 are achieved without saturation. The sample should then be rotated to its least X-ray dense orientation (normally at 90° to the most X-ray dense orientation), and the projection should be inspected to ensure there is enough contrast. It is advisable to then rotate the sample 360° to check the histogram and positioning again.

Note: Many scanners also allow the user to control the power, rather than the current. In this case, changing voltage (which will automatically adjust current to keep power unchanged) allows the relative brightness of the projection to be maintained as the voltage (thus penetration) is altered. This approach is best employed once the detector panel is approaching saturation in air spaces.

3.2.4.8 Running a Scan

Once a sample is positioned and appropriate scan settings determined, a scan can be started. The scan process is normally automated by the acquisition software. The sample may be rotated 360° to ensure it stays in the field of view. Subsequently – and universal to all scanners and scanning technologies in microtomography – calibration images are collected. These images pick up and correct for variations in the sensitivity of detector panel CCDs, which could otherwise create ring and other artefacts (Section 3.2.9). The majority of scanners possess linear detectors which require only two calibration images: a black or **dark field** (no X-rays) and a white or **flat field** (X-rays on, but no sample in the field of view). As with projections, calibration images are affected by noise, and thus, there is usually a frame averaging and/or exposure time option. Two minutes cumulative exposure time is usually sufficient for each calibration image. Most lab systems allow the user to set the number of frames to average and will automatically acquire (and eventually apply) calibrations. Some scanners automatically move the sample out of the field of view for the flat-field calibration, while other scanners require manual sample removal, with or without a facility to automatically return it to its previous position post-calibration. Failure to remove the sample during the collection of the flat-field calibration image will effectively ruin the scan data. If numerous objects of similar size and density require scanning, calibrations do not need to be re-collected for subsequent scans provided they employ the same settings. Calibrations are valid for about two hours in lab conditions, after which detector panel temperature changes can alter CCD sensitivities.

The acquisition software will often ask the user to enter the filtration or source details, so that these can be recorded with the scan data. It will also normally require the specification of a dataset name – we suggest this is descriptive and should feature, where possible, specimen numbers and the user's name to guard against data loss through confusions over ownership. Projections acquired by the scanner for reconstruction are sometimes sent via a network from the scanner's acquisition PC to an independent reconstruction work station, or just collected to be manually transferred and reconstructed elsewhere. Some systems allow the user to crop around a specimen at this stage to reduce reconstructed file sizes. However, this cropping does not always take rotation into account, so if employed, it should be conservative (i.e. not too close to the specimen).

After all data described earlier has been provided, the scan is completed without further user intervention. CT scanners are relatively reliable pieces of equipment (and a technician or lab member should have warmed up the source prior to use); however, occasionally, X-rays can cut out mid-scan, especially on older machines or those which have issues keeping the target under vacuum. This will necessitate a restart of the scan. Other common acquisition issues can arise network connection problems resulting in dropped frames, a lack of storage space on the reconstruction PC and – on high-resolution scans – sample movement or beam shift (drift of the activation volume on the target).

3.2.5 Medical Scanners

As with XMT, medical CT comes in a number of flavours. Medical tomographs are unlikely to be found in settings where palaeontologists will be expected to perform their own scans – this section hence provides an overview and important considerations, in contrast to the practical guide of XMT (Section 3.2.4). Common to all medical CT systems is the gantry. This holds the X-ray source and usually also the detector array, although in some old systems this is a complete but static ring. Immediately after the source, X-rays are passed through a collimator – a device that defines the X-ray field size/shape. Older systems typically possess a fan beam – that is, one with little depth and X-rays that diverge from a single point. In such systems, the minimum effective slice thickness is determined by X-ray beam width (and thus, ultimately, the collimator). This is coupled with a detector comprising a single row of photodiodes (in the z-axis) forming an arc in the xy-axis (an arrangement sometimes referred to as *SSCT*, single-slice spiral computed tomography). Modern systems typically have been 4 and 256 detector rows in the z-axis, allowing multiple slices to be collected simultaneously (an innovation driven by the requirement for fast scan times in a diagnostic setting). This *MSCT* (multi-slice spiral computed tomography, or multi-detector CT/MDCT) necessitates a cone-beam source which fans in both the xy- and z-axes: each detector element requires similar source strength to avoid artefacts. The collimator determines the geometry of the cone beam. Historically, diagnostic CT was a 2D technique resulting in a notable disparity between the xy- (in-plane) and z- (cross-plane) resolutions. However, MSCT has, in recent years, facilitated isotropic and sub-millimetre data. The cone-beam geometry has also necessitated improvements in reconstruction algorithms to correct the resulting artefacts.

3.2.5.1 Scanning Parameters

Modern medical scanners are spiral (helical) scanners: the sample is placed on a mobile mount/table and gradually moved in the z-direction while the *gantry* rotates (Figure 3.8). Z-movement is typically 1–10 mm/s, and a 360° gantry rotation can take as little as 0.5 seconds. It creates a spiral path around the scanned object and thus volume of data, in contrast to traditional 'step-and-shoot CT' where data can be collected at the z-position of each slice. The technique thus requires z-axis interpolation prior to tomographic reconstruction: that is, transaxial planar images necessitate estimation of projection data for the plane from adjacent (helical) data. Accordingly, the slice thickness can be arbitrarily defined, but the minimum thickness

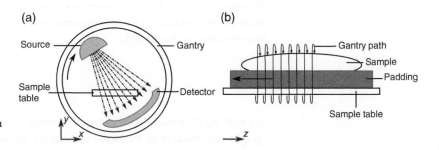

Figure 3.8 (a) A schematic of a spiral CT scanner in the xy plane and (b) its operation over time in the z-direction.

remains limited by the collimator and detector size. Medical CT literature will often refer to pitch – the ratio of z-axis movement in a single rotation to the slice thickness. As minimizing radiation dose is not a consideration for fossils, a pitch of 1 (a contiguous spiral) or smaller (an overlapping one) provides the best coverage. As with XMT, spatial resolution depends on numerous elements of the set-up, the image acquisition and the reconstruction algorithm (Section 3.2.4.1). Factors include the number and size of detectors, the field of view (determined by the fan beam angle), the focal spot size and energy, the magnification factor and the helical reconstruction algorithm (further discussion can be found in Bushberg et al. 2011). Field of view is defined in spiral CT as the size of the spatial region covered by a complete set of projection measurements – smaller fields of view will introduce fewer artefacts (Seet et al. 2009), allow for better resolution and, in some systems, can allow more projections to be collected. However, as with XMT, elements of the sample outside the field of view can produce artefacts. Voxel sizes of between 0.3 and 3 mm are typical of palaeontological studies with medical CT published in the last 5 years.

3.2.5.2 Running a Scan

Medical systems are typically available in medical or veterinary facilities, and radiographers are normally available to perform the scan. Little sample preparation is required; the fossil is positioned (with underlying padding) on the table, the radiographers and users leave the room, and a CT scan is conducted. Staff members of the institution can advise on a scan's pitch and adjust the field of view (which should be a good match for specimen size), voltage and current. Similar principles to XMT are required (Section 3.2.4.6). In spiral CT, the selection of these factors is best enabled by performing multiple short scans and iteratively improving the settings. With correct settings, calibrations can be acquired, and a scan performed. Data from medical scanners is often provided in a DICOM format – these can be loaded into, for example, the free software ImageJ and converted to whatever image format is required for visualization (Chapter 5).

3.2.6 Lab-Based Nanotomography (Nano-CT)

Scientists, palaeontologists included, have historically shown a tendency to use synchrotron sources for high-resolution scanning. However, in many cases, lab-based nanotomography set-ups provide similar resolution with greater convenience and economy. Nano-CT is a recently developed and largely experimental technique – only two companies currently make standalone nanotomography tomographs: Xradia (www.xradia.com) and SkyScan now Bruker (www.skyscan.be). As such, any scanning required will likely be conducted with a technician, and many details are similar to micro-CT. This section is intended to provide an overview.

3.2.6.1 Scanning Set-Ups

Lab-based nano-CT systems are varied, but akin in overall form and work-flow to microtomography scanners. The conventional XMT set-up is known as projection imaging in nano-CT parlance. At the nanometre scale, two factors impact on absolute resolution – source spot size, as in XMT, and the capture system (Withers 2007). Lab-based nano-CT systems possess a fine-focus cone beam (~0.4 μm spot size is achievable in some systems), much like that of lower-resolution systems, often created with a tungsten or molybdenum transmission target, and a beryllium window to reduce the target's heat load. Such sources do require – in contrast with a synchrotron – a small source-to-sample distance and can be low intensity. Even with a sufficiently small spot size, scintillator properties and the diffraction of light limit resolution in projection imaging to ~0.3 μm. In nano-CT, a number of optics techniques are circumvent this limit; these are generally lens systems that focus the source or magnify the scintillator. No standard form exists, although the most widely reported (largely still experimental) lab set-ups employ **Fresnel Zone Plates** (FZP, a series of rings used to focus X-rays via diffraction). These can be placed in front of the sample (with a central stop to block direct X-rays); in this case, they refocus X-rays onto a sample. Alternatively, a reflective condenser optic can be employed for this purpose. The sample is generally mounted on a high-precision rotation stage, behind which is an imaging zone plate: an objective to create a magnified image of the sample on the detector (Figure 3.9a).

With FZP optics, the diffraction-limited spatial resolution can be in the region of 50 nm. The limiting factor in uptake of FZP lab systems is the limited brightness of lab sources – they are currently predominantly used at synchrotrons. Alternatives which are currently largely limited to synchro-trons, but could in theory be used for lab-based systems in the future, use curved mirrors or refractive lenses to focus the X-rays. These include systems with two perpendicularly aligned elliptical mirrors (known as Kirkpatrick–Baez optics; Figure 3.9b; Kinney and Nichols 1992), especially efficient with high-energy X-rays, but prone to distorting projections, and those based on flat crystals at angles which take advantage of a property called asymmetrical Bragg diffraction to provide magnification (Figure 3.9c; Stampanoni et al. 2003).

One ingenious and relatively new form of high-resolution CT has come through the development of add-on modules for scanning electron micro-scopes, allowing these to be used for X-ray projection microscopy and CT scans. In these systems, the electron beam usually used for microscopy is directed at a metal wedge which becomes a reflection target. A custom-built stage allows a small sample to be mounted in the X-ray beam, and there is a detector panel added to the side of the microscope. Such systems can gener-ate energies up to 30 keV (but with a lower peak energy), but lack intensity.

Scanning at high resolution is awkward, especially in a lab setting. Problems include mechanical stability – even micrometres of wobble

(a)   **Nano-CT fresnel zone plate system**

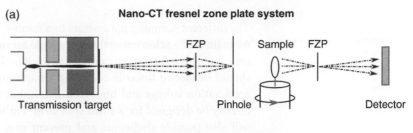

(b)   **Kirkpatrick-baez optics system**

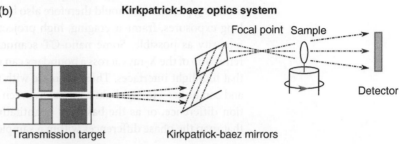

(c)   **Bragg multiplier system**

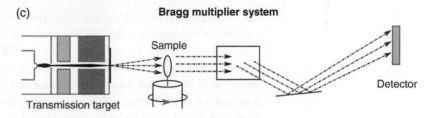

Figure 3.9 Three lab-based nano-CT systems – (a) Fresnel zone plate system which has limited applicability with lab sources. (b) Kirkpatrick–Baez optics system. (c) Bragg multiplied system.

between the source, specimen and detector can introduce artefacts and beam shift. This is especially true of SEM-based nano-CT systems where charging due to dirt on the target can deflect the beam. The low-intensity nature of small spot-size sources results in noise, and penetration depth is typically low, necessitating long acquisition times.

3.2.6.2   Sample Preparation

Sample preparation for high-resolution scans is challenging. Mounting specimens will often require sticking them to the head of a pin, or – for smaller samples – using FIB techniques (Section 2.3) to selectively mill around the region of interest, preparing 'matchsticks' of specimens (Lombardo et al. 2012). A micromanipulator can then be used to remove and mount the sample. For the highest resolutions, sample size will be in the region of micrometres to tens of micrometres. For the majority of palaeontological specimens (short of microstructure, histological or microfossil studies), this form of sample preparation will not be required. Some studies have reported a local tomography approach (Dierick et al. 2007; Dunlop et al. 2012) obviating the need for such sample preparation, but this remains experimental.

3.2.6.3 Scanning Parameters

The different scanning parameters in a nano-CT scanner will be an – often more limited – selection of those available for micro-CT. In plain projection imaging where the spot size provides a limit to the spatial resolution, settings should be geared towards obtaining a minimum spot size – such as a low acceleration voltage and high atomic number target (the target will likely already be designed for a small spot size). For most specimens, noise levels will also provide challenges and prevent easy reconstruction of the data. Scanning parameters should therefore also be chosen to improve the SNR – long exposures, frame averaging, high projection numbers and as high an intensity as possible. Some nano-CT scanners produce phase contrast – refraction of the X-rays across a boundary can result in dark and light fringes that highlight interfaces. These increase with the sample–detector distance and can be used to highlight contrast between materials with a low attenuation difference, or as the basis for quantitative reconstruction techniques that map the phase differences within a sample (Section 3.2.10).

3.2.6.4 Running a Scan

Once a sample is mounted, the scanning workflow is similar to that of micro-CT. Sample holders often slot into the centre of the rotation stage or can be tightened into a chuck. However, in contrast to XMT, finding a sample is often challenging; if the field of view is micrometres across, the sample will rarely immediately be situated between the source and the detector. At this resolution, almost all machines will have an adjustable sample stage above the rotation stage, allowing limited (~2 mm) travel (in *x*-, *y*- and *z*-directions) via joystick controls, software controls or thumb screws. These should be adjusted to bring the sample into view, which can then be rotated by 90° and the corrections repeated until the sample lies within the field of view for a 360° rotation. If the contrast is too low to be able to see the specimen, a ~30 μm gold particle can be placed on, and used to track, the sample. This can be achieved with the hair of a paintbrush. Once a sample is located and positioned, scanning can proceed as with traditional micro-CT, and problems such as sample drift corrected prior to reconstruction. Acquisition time may be in the region of hours to days, rather than the minutes to hours associated with micro-CT.

3.2.7 Synchrotron Tomography

Several synchrotron light sources exist which can be used for tomographic imaging of fossils. These facilities are powerful tools due to the high intensity of their X-rays, which allow (often rapid) acquisition of scans at high spatial resolutions. Moreover, their **monochromatic** (single energy) nature and high flux allows advanced reconstruction techniques (Sections 3.2.10 and 3.2.13) and removes issues with beam hardening. At very high resolution, **synchrotron tomography** is referred to as **synchrotron radiation X-ray tomographic microscopy (SRXTM)**.

3.2.7.1 Scanning Set-Ups

Synchrotrons vary, but all comprise a particle source (such as a Cockcroft–Walton generator) which functions as an electron gun. This is similar to a cathode ray tube – it bunches electrons, which travel through a linear accelerator (linac) to several hundred MeV. The electrons are then accelerated further in a small synchrotron (the booster ring) to an energy of several GeV, at which point they are injected into the storage (main) ring. This can be several hundred metres in circumference. The electrons circle within a vacuum, their path controlled by electromagnets which focus and bend the electron beam, occasionally being topped up from the booster synchrotron. Radiation is created through two mechanisms: bending magnets and insertion devices. Bending magnets, used to keep the electrons in their circular path, emit radiation of a continuous and wide spectrum (i.e. not just X-rays). They are less well focused and have lower brilliance than insertion devices (**brilliance** being a measure of beam quality based on the number of photons emitted per second, the beam's collimation, the source area and the spectral distribution). Insertion devices are found in the straight sections between bending magnets and come in two forms. Both use magnets to force electrons to oscillate in the horizontal plane as they pass through the device. The light from *undulators* is very bright and is emitted in a narrow beam of tuneable energy (or a continuous frequency range, if required). *Wigglers* comprise fewer magnets and emit a wide cone of light, with a broad X-ray spectrum. Insertion devices are capable of creating very high-energy X-rays and are the source for a number of tomography facilities, including those capable of scanning fossils. Experiments are conducted on beamlines – synchrotron light enters these through beam ports and passes through a series of optical devices on the way to an experimental hutch (Figure 3.10). These will typically include slits to limit the size of the beam, attenuators that absorb low-energy X-rays and reduce the heat load on the other optics, focusing mirrors and a monochromator, which creates a monochromatic beam via diffraction through two crystals. The angle of these crystals controls the energy of the beam, which is thus adjustable. The monochromator is followed by more slits and then a shutter which can

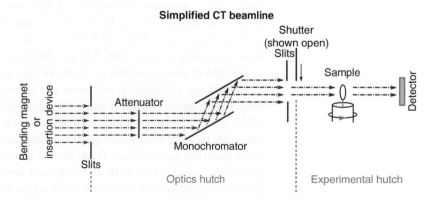

Figure 3.10 A schematic, simplified synchrotron beamline, based on I12 at Diamond Light Source, UK.

isolate the beam from the experimental hutch. Following these are typically a sample stage, scintillator, further optics to magnify projections and then a CCD that detects the signal. Beamlines run continuously (although down-time is periodically scheduled for maintenance), and each is designed for a particular kind of research, being optimized for different energies, sample sizes and X-ray techniques, for example. To be utility for palaeontological specimens, a tomography beamline needs to produce hard X-rays. Suitable tomography beamlines include ID19 at ESRF (European Synchrotron Radiation Facility), Grenoble, France; I12 at Diamond, Oxford, UK; and TOMCAT at SLS (Swiss Light Source), Villigen, Switzerland. These differ in energy range, field of view, resolution and the application procedure for beamtime. The majority of beamlines employ a competitive application pro-cedure, with projects awarded time based on the quality of their science, but provide beamtime for free if an application is successful. All these details can be found on the websites of the synchrotrons in question, and the capabili-ties of the beamline are generally well demonstrated in papers documenting their work. SLS is best known for true SRXTM, while ESRF and Diamond can also perform lower-resolution hand-specimen-scale tomography. Sample preparation is the same as outlined for XMT (Section 3.2.4.3).

3.2.7.2 Scanning Parameters

If an inexperienced user is awarded beamtime, they will be assigned a beam-line scientist who can provide advice and training in all aspects of the use of the beamline. The controls, protocols and nature of beamlines vary greatly – this section provides an overview of the issues and considerations required when synchrotron scanning. Sample preparation is generally similar to other forms of micro-CT, although due to the strength of phase-contrast effects created by a monochromatic beam (Figure 3.11), scanning a sample within sand or another granular medium can cause severe artefacts, even if the medium is entirely out of the field of view. Synchrotron detectors are often landscape in orientation, which means that many specimens will require scanning in horizontal sections and then stitching together verti-cally for the maximum possible resolution. A synchrotron beam is parallel rather than a cone; there is hence no geometric magnification available, and voxel size is constant irrespective of the source-to-specimen distance. Instead, resolution is altered by the optics – most beamlines will have multi-ple modules, each of which will have its own scintillators and optics and will allow for a range of voxel sizes. The parallel beam also limits specimen size, as the field of view is limited to the beam size. Creating large synchrotron beams remains challenging, and beam size is generally a few centimetres at most (ESRF has the greatest width of tomography beamlines at 60 mm). While synchrotron tomography is largely non-destructive, some beamlines – especially those of especially high flux or brilliance – can alter samples. In some materials such as glass, it can lead to discolouration. This could be a particular consideration with amber-hosted fossils; while no studies have been conducted on the impact of synchrotron radiation on amber, which

No phase contrast

Phase contrast

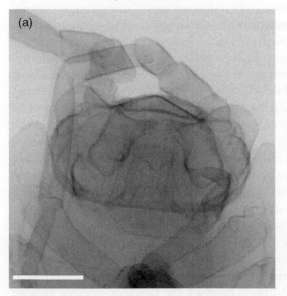 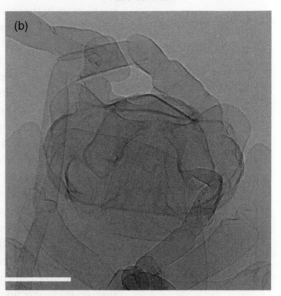

Figure 3.11 Projection images of a terrestrial isopod crustacean demonstrating plain attenuation contrast (a) compared with phase-enhanced attenuation radiographs (b). Both taken on an Xradia MicroXCT at the Manchester X-ray Imaging Facility, source at 60 kV, 10 W, attenuation image with a 4.2 mm sample to detector distance, phase contrast with 112.7 mm sample to detector distance. Scale bar 0.5mm. Images courtesy of Robert Bradley, University of Manchester.

appears unaffected by micro-CT, darkening is a theoretical risk in non-opaque fossils. Synchrotron radiation can also lead to the reduction of copper and thus impact on subsequent chemical analyses (Yang et al. 2011).

3.2.7.3 Running a Scan

Synchrotrons differ in their operating procedures. All have a shielded experimental beam hutch in which the sample stage and detector assembly are housed. This will be interlocked, preventing accidental opening of the shutters, and large enough for several users. As a safety precaution, synchrotron institutions enforce search procedures to ensure that nobody is present in the hutch before interlocks can be reset, and the shutter can be opened allowing X-rays into the hutch. Compliance is verified with keys or buttons inside the room and is of utmost importance as a lethal radiation dose from a synchrotron beam can accumulate in milliseconds.

Samples are often mounted on a screw-driven positioning goniometer for high-resolution beamlines or manually for hand specimens (numerous holders can generally be screwed onto the sample stage). A laser is sometimes used to mark the path of the beam, allowing the sample to be positioned correctly, as X-rays cannot be allowed to enter while researchers are positioning specimens. With the shutter open and X-rays on, the projections' magnification and field of view can be adjusted by changing the optics (usually an automated procedure), and a sample stage above the COR allows remote positioning of the specimen via beamline software. Other scanning parameters and protocols are similar to micro-CT: energy can, in some synchrotrons, be changed directly to the desired keV, impacting penetration and grey levels. With the ability to fine-tune a monochromatic beam, scintillator sensitivity becomes more important – any given material will be more sensitive at particular energies, and this can further impact the choice of beam energy. Exposure times can be altered, but for long exposures, brightness should be checked with the sample out of the field of view to ensure the flat field will not be overexposed. With some samples, the projection grey levels can be too low through a sample at all exposures below the point of detector saturation. Because filtration will not have the desired effect on monochromatic beams, in these circumstances, it is possible, but not ideal, to take a flat field with a different exposure time from projections. As with micro-CT, the grey levels through the thickest parts of the specimen should be high enough to indicate some penetration by the X-rays, but poor penetration is less of an issue due to limited beam hardening. The lack of filtration is generally accommodated for by the tuneable energy of the monochromatic beam. If phase-propagation imaging (Section 3.2.10) is required, the sample-to-detector distance should be tuned for phase retrieval (Weitkamp and Haas 2011). User-definable parameters include the number of projections (ideal values are usually between 1500 and 3000), the amount of rotation (in synchrotrons 180° or 360° are standard), exposure time and frame averaging. Scan times will generally be somewhere between 5 minutes and 3 hours.

3.2.8   Tomographic Reconstruction

3.2.8.1   Filtered Back Projection

Projections captured during scanning provide the basis for tomographic reconstruction, the process by which the tomograms are computed. Note that this is often simply termed reconstruction, a term also applied to digital visualization and hence a potential source of confusion. Projections, being radiographs, contain no depth information: rather, they represent an integration of the attenuation coefficient along the propagation path of the beam. To recover the attenuation map for each slice, an algorithm called **filtered back projection** is normally used. The maths of this process is based on Radon's theorem (Radon 1917), and in-depth explanations of the process abound in the literature (Natterer and Ritman 2002; Buzug 2008). We present here a non-technical summary.

To facilitate reconstruction, *sinograms* are created (Figure 3.12) – for each line of CCDs in the detector panel (which, when reconstructed, will constitute a single-slice image), projections are stacked, with time progressing from top to bottom (Betz et al. 2007). These are so called because any single point in the scanned object will trace a sinusoidal curve. While most reconstruction software treats this stage of the reconstruction process as inner working, the sinogram provides a valuable indication of issues with data acquisition. Vertical features represent issues or variability in any given

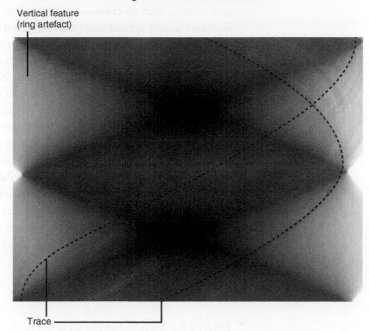

Sinogram for 180° rotation scan

Vertical feature
(ring artefact)

Trace

Figure 3.12 A sinogram for a 180° scan with two traces (i.e. the path of a single part of the object) marked on. Also visible are vertical features which will, following reconstruction, form ring artefacts.

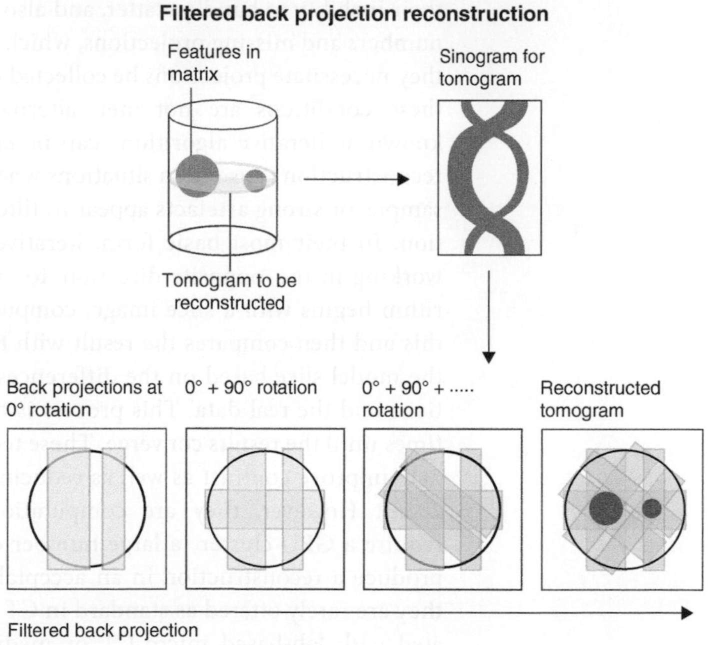

Filtered back projection reconstruction

Features in matrix

Sinogram for tomogram

Tomogram to be reconstructed

Back projections at 0° rotation

0° + 90° rotation

0° + 90° + ... rotation

Reconstructed tomogram

Filtered back projection

Figure 3.13 The filtered back projection algorithm. The process of creating a reconstructed cross section is shown for a single tomogram: first the creation of a sinogram, which will include applying calibrations, and then filtered back projection of the data.

detector channel (Figure 3.12). Horizontal lines or features will be temporal in nature, relating to source variations over the course of a scan. To create a reconstruction from the sinogram, each line is successively superimposed over a square grid at the angle of its acquisition (Figure 3.13). This back projection is akin to 'smearing' the projection across the reconstructed slice and creates blurred cross-sectional images – for example, a single point becomes widely spread and possesses a surrounding halo. To mitigate this effect, mathematical filters are applied prior to reconstruction. A simple example is a ramp (high-pass) filter, which removes low-frequency signals, but passes or amplifies high ones (Herzog 2002). Because sharp features are high frequency, but blurring is low and thus minimized, this kind of filtering can provide detailed and sharp reconstructions. It does, however, amplify noise and is thus usually combined with low-pass filters which allow the low frequencies to be retained and block higher ones. Further filters can then be applied to enhance the signal. In spiral medical CT scanners, filtered back projection will be preceded by the interpolation step (Section 3.2.5), and in all forms of CT, numerous filters exist that emphasize different aspects of the data in reconstructions. Traditionally, slices map X-ray attenuation at right angles to the axis of rotation, but datasets can easily be modified to form image stacks at arbitrary angles (with the risk of introducing artefacts) or just treated as a volume (Chapter 5).

3.2.8.2 Iterative Reconstruction

Filtered back projection is an **analytical reconstruction method** – these elegant algorithms are computationally efficient, but have limitations in their inability to handle scatter, and also deal poorly with low projection numbers and missing projections, which result in artefacts. Furthermore, they necessitate projections be collected over a minimum of 180°. Where these conditions are not met, alternative reconstruction techniques known as iterative algorithms can be employed. Additionally, iterative reconstruction is useful in situations where there is poor penetration of a sample, or strong artefacts appear in filtered back projection reconstruction. In their most basic form, iterative algorithms can be thought of working in the 'opposite direction' to analytical techniques – the algorithm begins with a slice image, computes theoretical projections from this and then compares the result with true projections. It then updates the model slice based on the differences between the computed projections and the real data. This process is then repeated a large number of times until the results converge. These techniques are versatile – they can help improve contrast as well as reducing artefacts (see, e.g. Hsieh et al. 2013). However, they are computationally expensive and normally require a GPU cluster (a large number of graphics processing units) to produce a reconstruction in an acceptable amount of time. At present, they are rarely offered as standard in CT reconstruction packages associated with lab-based micro-CT or medical systems, but as computing power increases and algorithms improve, iterative techniques are likely to become increasingly important. In view of the difficulty inherent in scanning fossils, and the associated artefacts, this could have a large impact on tomography in palaeontology.

3.2.8.3 Laminography

Other experimental techniques exist to facilitate the scanning of unusual geometries and with atypical parameters. One of the most promising of these is computed **laminography**. Many palaeontological specimens are highly anisotropic in that they are flattened (slablike), making penetrating the long axis difficult without saturating the detector panel during the flat field. Laminography attempts to overcome this variation in beam transmission by using a rotation axis that is inclined at less than 90°, rather than being perpendicular to the incoming beam as in traditional CT. Otherwise, scanning set-up is the same as in micro-CT. This facilitates good transmission at all angles and is especially well suited to synchrotrons due to the monochromatic, high flux and highly collimated beam. Micrometre-scale resolution can be achieved with this technique (Helfen et al. 2006); however, it does introduce some artefacts. The technique has been applied with some success to fossils (Houssaye et al. 2011). Further development and more widespread application could revolutionize studies of some of the more challenging anisotropic fossils.

3.2.8.4 Running a Reconstruction

Reconstruction workflow varies markedly between systems and institutions, but it is rare for a user to become deeply involved with the process – unless using experimental techniques or working on a new synchrotron beamline where integrated software has yet to be coded. Rather, scanning systems each have their own software workflow allowing the user to conduct reconstruction in a (hopefully) intuitive fashion and typically hiding some fully automated stages. In lab settings, reconstruction software will often initially offer an option to view the first and last projections to check for sample movement or beam shift during the scan (assuming 360° rotation, or using a reflected image for 180°). Either of these – if too large – will result in blurred and unusable datasets. Computational reconstruction then normally starts with the application of corrections derived from dark- and flat-field images to the projections, followed by the building of sinograms.

Most reconstruction workflows will then help the user find the *COR*, a requirement for most reconstruction algorithms. An incorrectly placed COR results in a loss of resolution, blurring and streaks due to misregistration of the projection data. COR is usually measured in pixels from the left of the detector panel and should ideally be aligned with the centre of the detector. There is, however, often a small element of drift, or deliberate offset for operational reasons, or to avoid the worst ring artefacts. Some systems do not allow *x*-axis movement – in these cases, the COR does not vary, and this step is omitted. Software differs – some packages define the COR with the click of a button and an automated search procedure. In many others (or if automated COR fails), a number of different CORs will be created for different positions, and the user will be required to select that closest to the true COR. This will be the image with the least streaks and the best-defined edges. A smaller range of CORs can then be reconstructed, and the process repeated until the COR has been correctly identified to within a pixel. This entire process is often conducted on an upper and lower plane to allow for the correction any tilt in the COR.

With COR determined, the data can be reconstructed. Some systems provide the option at this stage to crop parts of the volume, allowing for smaller reconstructed datasets and faster reconstruction times. This can be thought of as performing the region of interest specification of (Section 5.3.1) at an earlier stage, avoiding wasting time reconstructing voxels which will be discarded. It is recommended if possible, although inspection of multiple preview slices is recommended to avoid truncating the specimen. Dataset reconstruction can take from ten minutes to several hours depending on the size of the dataset, computational resources and algorithm used. In micro-CT labs, this workflow might be conducted on a reconstruction PC – that to which the scanner sends its projections. Each tomograph manufacturer will generally have proprietary reconstruction software tuned to its own system geometry (this may be read from an XML file found with the projections). At synchrotrons, there is often a GPU cluster that allows reconstructions to be processed in parallel – however, due to larger datasets resulting from the higher specification detector cameras, this is often not noticeably quicker than reconstruction in a lab-based setting.

3.2.9 Artefacts

Many artefacts can detrimentally affect CT data. This is especially true of fossils, which are particularly difficult to scan because they are dense, resulting in low transmission and high noise; often anisotropic; and rarely pristine in their preservation. Artefacts can also result from the **polychromatic** radiation of laboratory sources, finite resolution and X-ray scatter. These can all alter the grey levels in a tomogram (Figure 3.14) and can obscure – or worse, be mistaken for – genuine parts of a fossil's morphology. Numerous techniques have been developed to overcome these, some of which will be outlined in this chapter. More in-depth overviews can be found in Davis and Elliott (2006), Barrett and Keat (2004) and Boas and Fleischmann (2012).

3.2.9.1 Noise

Noise is inherent to even the cleanest CT scan. There are numerous different forms of noise – *statistical* (otherwise known as quantum) noise will usually appear as mottling in the slices, making small features harder to see and creating a rough surface texture in 3D reconstructions. It can also make thresholding slice images to produce an isosurface problematic (Section 5.5.2.4). Statistical noise occurs because X-rays are transmitted in a finite number of quanta – mottling results from the statistical fluctuations counting these (Figure 3.14). The effect is often more notable in fossils because of the low SNR in many scans of palaeontological material. This noise has

CT Artefacts

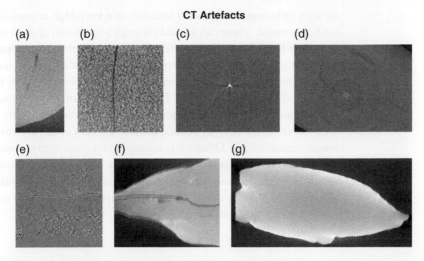

Figure 3.14 Different artefacts typical of CT scans. All scans are siderite nodules. (a, b) Statistical noise is shown within a large nodule at two magnifications. (c) Streak artefacts surround a high-density material (most likely pyrite). (d) Ring artefacts resulting from differing sensitivities in zones of the detector panel. (e) COR artefacts are expressed as arcs around voids and cracks in the rock. (f) Sample movement causing doubling of a crack and an edge in the sample. (g) Beam hardening in a large siderite nodule. Also visible are subtle streaks on the long axis. All projections have had the contrast altered to highlight the artefacts.

greater impact at higher resolution. It tends to be more pronounced in the centre of a scanned sample due to lower transmission in this region and can be anisotropic in objects with a high aspect ratio (i.e. more prevalent in one orientation). More advanced reconstruction software will often offer a noise reduction option during reconstruction. If this is insufficient, it is best countered by repeating a scan with higher energy to gain better transmission (Davis and Elliott (2006) recommend a minimum of 16% X-ray transmission through the specimen centre), longer exposures to increase the SNR, more projections or frame averaging. If a re-scan is impossible, the impact of noise can be minimized at the cost of resolution by downsampling (**binning** – Section 5.5.2.4), through **iterative reconstruction algorithms,** or by applying a median or Gaussian filter to the reconstructed data (see Brabant et al. 2011). The latter filters introduce blurring, however, and so can severely affect edge quality. This can be mitigated, to an extent, by more complex (non-linear) filters which apply smoothing preferentially to non-edge regions. This is particularly effective in synchrotron data which tends to have sharp edges as a result of phase contrast.

3.2.9.2 Streak Artefacts

Straight edges can introduce streak artefacts most noticeably in scans with low projection numbers. These dark and light streaks emanate from, and run sub-parallel to, straight edges (Figure 3.14). Similar artefacts can often be seen radiating from a small instance of a very high attenuation material such as a metal. These can extend through a specimen and are worsened by beam hardening. Streaks can rarely be entirely removed, but repeating scans with an increased number of projections reduces their prominence. Ideally, the minimum projection number should be the pixel width of the detector multiplied by $\pi/2$ (Davis and Elliott 2006). If an increase in projection number and the associated longer scan time is impractical, more projections with lower exposures may reduce the artefact (but could increase noise and lower transmission). Otherwise, manual correction during visualization is the only solution. Streaks are particularly common and problematic in fossils with fissile host rocks where the straight cleavage results in streak artefacts obliterating detail on the rock surface. This can be overcome by scanning in another medium such as sand.

3.2.9.3 Ring Artefacts

Ring artefacts are circular light and dark rings around, and most severe at, the COR of a scan (Figure 3.14, see also Figure 3.12). They result from variable sensitivities in detector elements – each pixel, when reconstructed, will create a circular trace in the resulting slice image (or a vertical feature in the sinogram). It is these varying sensitivities for which the flat-field/light calibration corrects. However, it may not do so fully if the calibration is too old, as detector sensitivity can change with time and temperature. If rings are an issue, the scan should be repeated with a longer calibration. Some systems bypass this problem entirely by moving the specimen or detector a small amount in the x-direction between each projection: this facility will be automated, if present.

3.2.9.4 COR and Motion Artefacts

An incorrect COR will result in streaks, apparent broken edges (which are, in reality, continuous) and double-edged slice images. In a COR artefact, the double edges will be a constant distance apart (although if they get larger from the centre of the image, this could be due to reconstruction with an incorrect cone angle). Small bright objects will form an arc-shaped smear (Figure 3.14; although these can also be caused by specific, high-contrast, particle morphologies). These artefacts can be corrected by re-calculating the COR. This error is superficially similar to motion artefacts caused by sample movement or beam shift (or very rarely detector movement). This will also result in double edges in slice images, but ones which vary in thickness. Often when manually finding a COR, sample movement will result in areas of a slice having different apparent CORs. If this has occurred, the specimen should be re-scanned more securely mounted or at a lower magnification (small movements will have a greater relative effect the higher the zoom). Sometimes the motion can be caused by parts of the specimen catching on objects in the scanner: a 360° rotation, observed through the window in the door, should make it apparent if this is the case. Post-scanning correction is possible, but is rarely easy.

3.2.9.5   Beam Hardening

Beam hardening is one of the most common artefacts when scanning fossil material with lab-based or medical scanners (Figure 3.14). It results from the usually high density and large size of fossils. The artefact is caused by the polychromatic nature of the X-ray sources in these systems (and, accordingly, is not an issue on a monochromatic beam, which has a single energy of X-rays). Lower-energy X-rays are preferentially attenuated through a material, an effect most noticeable in dense objects. In any given projection, this attenuation is greater through the centre of a specimen than its periphery. In slice images, this creates an effect with a lighter halo towards the exterior of an object and an apparently less dense zone in the centre. This causes problems when thresholding (Section 5.3.4.2) – the brightness change due to beam hardening can often exceed the contrast between fossil and matrix. Furthermore, the artefact obscures details towards the centre of a sample. Beam hardening can be overcome in the first instance by scanning with a thicker filter (or adding one if none were previously used) or changing the metal in the filter. This could either absorb a larger proportion of low-energy X-rays or create a near monochromatic beam, as previously described, and will often be coupled with an increased exposure time to compensate for lower intensity. Beam hardening is unavoidable in some fossils, due to their sheer size and the power limit of lab-based sources. If this is the case, beam-hardening correction is an option in a number of pieces of reconstruction software. This will generally linearize data with a theoretically derived calibration curve or, in some systems, allow one to be experimentally collected. For palaeontological applications, correcting scanning parameters is usually more effective than post-acquisition correction, which assumes a smaller number of phases than are present in most fossils and a more regular shape.

3.2.9.6   Cone Beam and Field of View Artefacts

Artefacts related to the nature of the cone-beam geometry in micro-CT and multi-slice medical scanners also exist. Near the top and bottom of the detector panel, slices will correspond to X-rays which have not travelled a flat plane to reach the detector. This introduces slight blurring, but is rarely an issue in fossils. Cone-beam artefacts can be avoided, if necessary, by reducing the cone angle or positioning regions of interest closer to the horizontal midplane of the cone. Some reconstruction algorithms with large elements outside the field of view can produce bright pixels in the slices towards the edge of the field of view. These can be overcome by reducing the zoom on a scan.

3.2.9.7   Partial Volume Averaging

Partial volume averaging is inherent to all CT data. It results from the fact that the intensity of any given voxel in a reconstruction is proportional to the mean attenuation coefficient within that cube; if different materials exist within the same voxel, the grey level will be a mean of the attenuation of all materials. As a result, at material interfaces in slices, there is usually a gradient, which appears as blurring (Abel et al. 2012). The closer an anatomical element is to the voxel size, the more pronounced this effect will be. If a sample is scanned at the maximum magnification possible on any CT system, the only means by which partial volume averaging can be mitigated is by re-scanning on a scanner with an alternative detector with greater dimensions.

3.2.10   *Phase-Contrast Tomography*

In some fossils, irrespective of the scanning parameters or system chosen, it remains impossible to differentiate fossil from matrix using attenuation-based reconstruction. The root cause is a lack in attenuation contrast between fossil and host rock – the two may be of identical composition or very similar densities. When this is the case, phase-contrast tomography can provide superior results: a parallel technique, phase-contrast microscopy, is used with visible light to differentiate two substances of similar transparency but different refractive index. The majority of published work employing the technique has arisen from synchrotron work – particularly that of ESRF (e.g. Tafforeau et al. 2006). However, phase-contrast-based studies have also been conducted with lab-based scanners (Dunlop et al. 2012). The technique – especially on lab-based systems – remains experimental.

3.2.10.1   Phase Contrast

Phase contrast arises from the refraction of X-rays at material boundaries. While refraction is generally small, it can be measured very accurately (Als-Nielsen and McMorrow 2011). Because the passage of the radiation through a sample is delayed by differing times, the waves become out of phase (Figure 3.15). The change in a refracted beam's phase is proportional

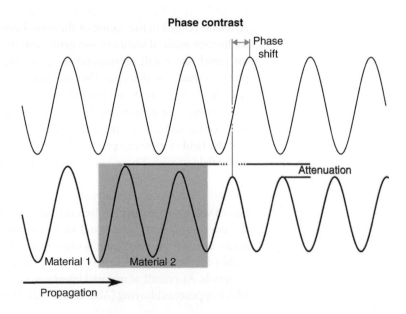

Figure 3.15 Phase shift and attenuation through differing materials.

to its angular deviation due to refraction and easier to measure. X-rays can be visualized as a wavefront – a plane with constant phase perpendicular to the propagation direction (like a wave on a beach, but perfectly straight; although note this is true of a monochromatic synchrotron [parallel] source, a lab source will be hemi-spherical). When this penetrates an object, changes in the material will distort the wavefront: refraction will change the propagation direction and phase, while interference will create intensity variations in the projection image (Wilkins et al. 1996). These distortions are recorded by the detector as bright and dark fringes around material interfaces in a sample. They appear in the majority of synchrotron and high-resolution lab scans and create similar fringes in slices and projections. The fringes become increasingly prominent with larger sample–detector (*propagation*) distance (although caveats apply, at long distances, fringes will vary sinusoidally, and multiple fringes appear with monochromatic sources). Phase-contrast-based tomography is possible because these effects are cumulative along the path of the X-rays: by transforming phase contrast into variations in intensity, it is possible to enhance the contrast between different materials. Thus, when employing quantitative phase retrieval, this technique in essence maps in three dimensions the distribution of the refractive index. Phase-contrast imaging comes in a number of different forms, which are introduced later.

3.2.10.2 X-Ray Interferometry

This is one of the oldest forms for phase-contrast imaging and relies on three silicon crystal wafers (Figure 3.16). The first diffracts and splits the monochromatic incident beam into two identical monochromatic

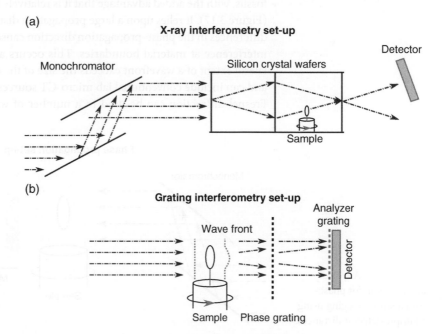

Figure 3.16 A diagram showing: (a) An X-ray interferometry scanning set-up. (b) The more recent grating interferometry technique which is possible with lab sources.

wavefronts. The second transmits these beams, but changes their direction making them converge. One – a reference beam – remains undisturbed, while the other passes through the sample. The two beams meet at the third crystal and create an interference pattern which can then be detected and is dependent on the difference between the two beams. Phase shift can then be calculated using phase-stepping or Fourier-transform methods (Keyriläinen et al. 2010). This can be applied to a rotating sample to create projections, and tomographic reconstruction can then be conducted. This method is very sensitive to phase changes, but the third crystal can limit resolution by introducing blurring, and the crystal arrangement creates a limited maximum field of view. A related technique, known as grating interferometry, has recently been demonstrated (Withers 2007). This employs two gratings, a phase and then an analyser grating, which create interference patterns on a detector. Comparison of this pattern with and without the sample placed in front of the grating pair allows the interference pattern of the sample to be retrieved and hence yields a phase-contrast map. This requires an intense source and a high-resolution detector, but could – in the future, with the development of suitably sized gratings – open up this technique to conventional sources. Palaeontological applications to date have been limited to imaging rather than tomography (Ando et al. 2000).

3.2.10.3 Phase-Propagation X-Ray Imaging

This form of phase contrast is otherwise known as propagation-based imaging, or in-line holography. It is produced by hard X-rays during transmission due to a process known as Fresnel diffraction (Mayo et al. 2012). Because it works with high-energy X-rays, it is particularly suitable for fossils, with the added advantage that it is relatively simple in terms of set-up (Figure 3.17). It relies upon a large propagation distance, which allows small differences in the phase-propagation direction caused by the sample to cause interference at material boundaries. This occurs as long as the size of the lateral extent of a wavefront exceeds the size of the elements being imaged – and can include conventional lab micro-CT sources (Jonas and Louis 2004). Fresnel diffraction can be used in a number of ways – quantitative phase

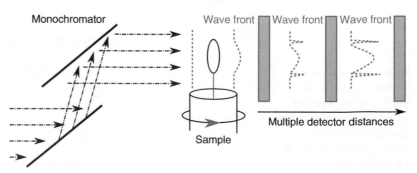

Figure 3.17 A typical arrangement for phase-propagation imaging using multiple detector distances.

retrieval is one of the most promising avenues of development. With monochromatic synchrotron sources, phase-contrast holotomography can be conducted; this requires numerous projections of an object to be taken at different distances from the detector (Cloetens et al. 1999). Different propagation distances allow the phase shift caused by transmission through the sample to be retrieved for each projection, from which point filtered back projection can be employed as standard. This is currently only regularly implemented at ESRF, but offers very promising results: the only weakness being that gradual variation of the refractive index within a specimen can be difficult to accurately recover. Many algorithms exist for phase retrieval: examples include both iterative and analytical approaches (Keyriläinen et al. 2010 and references therein). All differ and have their own limitations based upon the assumptions required by the algorithm – some require projections at a single propagation distance, speeding up acquisition time, but many impose restrictions such as sample composition, X-ray optical properties or the source nature. While phase-propagation imaging is one of the best techniques available for difficult palaeontological samples, details and procedures are not well established enough to be conducted without an experienced beamline scientist or similar guidance; we recommend that these experts are consulted on the best approach for any given specimen and concurrent scanning requirements. Propagation-based imaging is currently the most suitable technique for use with fossils. The vast majority of the work that has already been published has been from ESRF: a comprehensive overview and useful introduction are supplied by Tafforeau et al. (2006). We introduce one case study to demonstrate possible considerations (Section 3.2.12.3). Another pertinent example, which is likely to be increasingly employed at other beamlines due to its versatility, is the algorithm of Paganin et al. (2002), which requires a single propagation distance and can be applied with the freely available software (Weitkamp and Haas 2011). This makes it more widely available and accessible than many other approaches. When scanning, the most important element is the propagation distance, which depends on the pixel size and energy (Weitkamp and Haas 2011).

3.2.10.4 Analyser-Based Imaging

This approach to reconstruction using phase contrast comes in many forms, with a bewildering array of names, including refraction-contrast radiography/introscopy, phase-dispersion imaging/introscopy, diffraction imaging and diffraction-enhanced X-ray imaging (but see Keyriläinen et al. 2010 for a more exhaustive list). A typical set-up (Figure 3.18) will comprise a monochromatic (synchrotron) source, a sample and then an analyser crystal between the sample and the detector. When the transmitted beam hits the analyser, a very limited number of X-rays at specific incidence angles are reflected (due to a process known as Bragg diffraction). Rotating this crystal alters these angles, differentiating non-deviated rays, and those refracted or scattered by the sample. Recording projections at a number of different analyser angles allows a dataset to be created which can facilitate quantitative

Analyser-based set-up

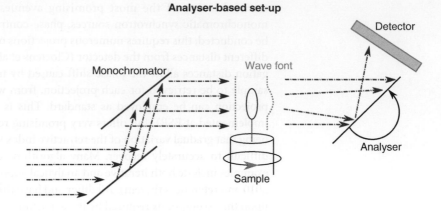

Figure 3.18 The arrangement for analyser-based imaging with rocking analyser crystal.

phase retrieval. Projections are collected through rotation at numerous angles of the analyser crystal to allow tomographic reconstruction. Limitations to this technique are the field of view, which is limited by the size of the analyser and the requirement of a parallel beam. Furthermore, blurring in the analyser crystal provides an absolute limit on the spatial resolution. It has recently been demonstrated that this technique is possible with a laboratory source (Muehleman et al. 2010); however, due to the more involved nature of the phase retrieval, it seems less likely to impact on palaeontological studies in the future than phase-propagation methods, despite a high sensitivity to changes in refractive index. Published studies employing analyser-based imaging are lacking.

3.2.11 Scanning Considerations

3.2.11.1 Choice of System

There are a number of factors to consider when choosing which CT technique to use for any given fossils. Often the choice will balance pragmatism and technique best suited to the sample; rarely are all techniques easily accessed or even available. CT has numerous advantages over many other techniques being a widely accessible, well-documented, non-destructive approach which is applicable to a very broad range of specimen types. Disadvantages are an upper size limit (sometimes due to the field of view, but otherwise due to the finite penetration of an X-ray source), a lack of attenuation contrast in some specimens, issues with small samples in large pieces of host rock and difficulties with highly anisotropic specimen morphologies. When a choice of systems is available, decisions will be based primarily on scale – each scanning technique has its own ideal size range, although these do overlap. From medical CT to micro-CT, to nano-CT, there is a drop in voxel size, coupled with an increasingly limited field of view. Synchrotron beamlines straddle micro-CT and nano-CT size ranges, but offer further advantages (Section 3.2.7). Above 200 mm, samples will most likely require a medical scanner. However, because medical scanners are designed to minimize patient exposure, they may struggle to penetrate larger specimens. If they lack the penetration power for a fossil, industrial CT systems exist and may be the only recourse. These often possess very powerful sources, but are challenging to access; little palaeontological work has been conducted on them, and examples in the literature are sparse. Below 2 mm sample size, a fossil will likely be too small for many lab-based micro-CT scanners (although this does vary – some SkyScan dental micro-CT scanners can successfully scan objects on the scale of 1 mm). Either SRXTM or lab-based nano-CT will normally be required for samples at these scales; the former is better established and will normally provide better scans, but beamtime is typically limited. Furthermore, both lab scanners and some synchrotron beamlines are designed for organic material and thus may lack the energy required, even for small fossils. For the majority of fossils of intermediate size (2–200 mm), micro-CT is usually preferable due to its accessibility and versatility in energy, scanning parameters and resolution. It is an ideal tool for most invertebrate fossils and smaller vertebrates, except where attenuation contrast between specimen and matrix is low or absent; this can generally only be ascertained by a test scan. Such difficult specimens may require synchrotron phase-contrast imaging.

3.2.11.2 Safety

Ionizing radiation is dangerous, but most lab scanners are inherently safe – that is, it is impossible to turn the X-rays on while the door is open because there are electrical contacts on the door to complete the circuit. Scanners will thus automatically cut off the source if the door is opened with the X-rays on (note that while this will do the user no harm, it could damage the machine). There are also a large number of interlocks on lab systems to ensure safe use. All labs will have safety training prior to unsupervised use of the equipment. Similarly, in synchrotron environments, there will be safety training and a search or hutch locking procedure to follow. We strongly advise against breaking safety rules for any reason.

3.2.12   *The Future: Three-Dimensional Elemental Mapping*

The CT revolution has demonstrated the importance and impact of CT as an approach to the study of 3D fossils. However, X-ray computed tomography is a rapidly developing field, and further advances in palaeontological applications are unlikely to be limited to refinements of established techniques. In particular, advances that will allow CT to map elemental distribution in three dimensions – and, further into the future, potentially the chemical state of elements – could revolutionize the study of the geochemistry and taphonomy of fossils and processes of preservation. Additionally, they would enhance contrast in hard-to-scan fossils and facilitate, in some cases, more automated reconstruction workflows (Section

3.2.12.3). Three such novel CT-based techniques with palaeontological potential are described here.

3.2.12.1 K-Edge Subtraction

When the energy of an X-ray source increases, there are a number of sudden increases in the attenuation coefficient at energies unique for any given element (Section 3.2.3.1). The largest – the K-edge – relates to the binding energy of the K-shell electron of an element and increases with atomic number. If scans are conducted at two slightly different energies that bracket an element's K-edge, the difference between attenuation in the two scans will be dominated by the K-edge signal from that element (Krug et al. 2006; Cooper et al. 2012). If this difference is used as a basis for reconstruction – that is, the sinograms or reconstructed individual tomograms are subtracted – it theoretically allows the non-destructive mapping of that element's distribution in three dimensions. To date, this technique has had relatively limited uptake (Ritman 2004), although it has been adopted in diagnostic settings. Current work is synchrotron based, taking advantage of the ability of monochromators to finely tune X-ray energy. Creating the required narrow spectral bandwidth with a lab-based source remains challenging. Studies are necessarily limited to mapping elements with a K-edge within the range of the beam's energy which would limit individual beamlines to a finite range of geological problems (TOMCAT: Cu to Nd, ESRF ID19 Mn upwards, Diamond I12 Ho upwards). Nevertheless, K-edge subtraction has great potential for palaeontological specimens being already established and the processing being relatively straightforward.

3.2.12.2 XANES Tomography

X-ray absorption near edge structure (XANES), otherwise known as near edge X-ray absorption fine structure (NEXAFS, a term more commonly used for lower-energy spectroscopy), is a well-established form of X-ray absorption spectroscopy when applied in two dimensions. It uses the structures in the vicinity of an absorption edge to deduce aspects of the chemistry of a sample. This will vary depending on the chemical environment of the probed element, creating a characteristic spectrum. Using the attenuation pattern of X-ray energies within 50–100 eV of an absorption edge, XANES can provide information about the oxidation state of the element of interest, for example. The same principle can be applied in three dimensions with the aid of a synchrotron – if a sample is repeatedly scanned at, for example, 2 eV energy steps, the reconstructions can be used to create a XANES spectrum for each voxel (Rau and Somogyi 2002; Rau et al. 2003). This can, in theory, provide more complex chemical maps than K-edge subtraction which include information of the chemical speciation of an element, for example, its oxidation state. However, the technique remains highly experimental, very difficult and time-consuming to perform, and no software exists to facilitate its application. In the future, with further development, if practical it could prove to be another incredibly valuable tool for taphonomic studies.

Another similarly experimental 3D chemical mapping technique is synchrotron energy-dispersive X-ray diffraction tomography/tomographic energy-dispersive diffraction imaging (Hall et al. 1998). This can non-destructively recover X-ray diffraction and fluorescence information in three dimensions, providing information about the elemental composition and related crystalline phases in a material.

3.2.12.3   Colour CT

Colour CT involves the coupling of a conventional (polychromatic) laboratory-based high-energy X-ray source with a specialized detector capable of resolving an X-ray spectrum for each pixel rather than a single intensity level (Jacques et al. 2013). In this study, the authors' 80×80, 250 µm pixel camera has allowed real-time imaging to very high X-ray energies. The camera, known as HEXITEC, comprises a data acquisition system and a 1 mm thick CdTe single crystal detector (20×20 mm). It possesses an energy resolution of ~800 eV at 59.5 keV and ~1.5 keV at 141 keV. Jacques et al. (2013) report that an experimental scan has successfully allowing both the spatial and spectral X-ray images (the latter binned into spectral bands). Accordingly, spectra for individual pixels can be extracted, as can images at individual energies. In the future, this approach could provide a lab-based means of performing K-edge subtraction, the easy and automated spatial mapping of different absorption edges and **XANES tomography** without a requirement for multiple scans. It also invites the use of multivariate statistical techniques such as principal component analysis (PCA) to find natural groupings within a sample-based spectral similarity, potentially identifying and automatically thresholding the different phases within a fossil. Colour CT remains highly experimental, scan acquisition times relatively long, and spectral detectors limited in size and sensitivity; nonetheless, from a palaeontological perspective, the potential for multiple approaches to elemental mapping combined with the use of a lab-based source renders this one of the most promising techniques in development.

3.2.13   Case Studies of Methodology

In order to provide exemplars of CT methodologies described earlier, we detail three different studies.

3.2.13.1   X-Ray Microtomography: Carboniferous Arthropods

Carboniferous arthropods are often preserved in the form of voids (sometimes kaolinite infilled) within siderite nodules, which range from 10 to 150 mm in size. These fossils are often found in association with coal mines, and thus, large collections were accumulated in the UK, the USA and Central Europe over the course of the industrial revolution. Published accounts have, however, normally been limited to the morphology visible on a cracked surface. The use of latex casts helps, but can still only recover

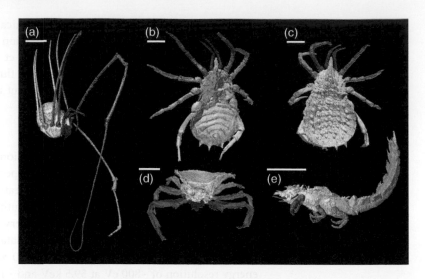

Figure 3.19 A range of siderite-hosted Carboniferous arthropods: (a) A harvestman. (b) Ventral and (c) dorsal views of a trigonotarbid arachnid. (d) Front view of a trigonotarbid arachnid. (e) A neopterous insect nymph. All scale bars 5 mm. (e) – *Source*: Garwood et al. (2012).

limited information. The 3D preservation and compositional/density contrast between fossil and void makes them ideal for analysis with micro-CT. However, due to the X-ray dense (iron-rich) nature of the nodules, successful scanning awaited the advent of XMT scanners capable of penetrating the nodules. Subsequently, it has been applied over a number of publications that have generated insights into the morphology, evolutionary relationships and palaeobiology of these organisms (Figure 3.19; Selden et al. 2008; Garwood et al. 2009; Garwood and Dunlop 2010; Garwood et al. 2011; Garwood and Dunlop 2011; Garwood et al. 2012; Legg et al. 2012).

Scans for this publication series were largely conducted on a Nikon (formerly Metris X-Tek) XT H 225 cabinet scanner at the Natural History Museum, London. Samples were mounted with the part and counterpart held together with elastic bands (for larger nodules) or within cling film (smaller nodules). These were mounted on blocks of florist's foam. Early attempts to scan part and counterpart separately and realign during digital visualization were unsuccessful and provided little improvement in the resolution. Nodules were mounted longest axis vertical, but were often sub-spherical, and thus, datasets were all acquired in a single scan; concatenating scans would have provided marginal improvement in resolution, but at the cost of doubling of (the already long) scan time, and complicating reconstruction. Specimens were manually centred on the manipulation stage, and then zoom was set. All scans employed a tungsten reflection target. Suitable projection grey levels were sought, initially with no filter, by altering the voltage and energy. If the maximum energy was not sufficient or increasing the exposure saturated the detector panel, a copper filter, starting at 0.1 mm up to a maximum of 2.5 mm, was added. For all but the largest nodules, suitable settings were possible (although on the medium axis of bigger specimens, the lower histogram levels were around 3000 on a 16-bit scale, 0–65,536). Typical settings for a 50 mm nodule were a 1 mm copper filter, 200 mA current and 225 kV acceleration voltage. 3142 projections, each with an exposure time of 0.3–2 seconds, were typical, resulting in 20–105 minute scans. Calibrations of a minimum of 2 minutes for flat and dark fields were collected as part of the acquisition, and projections were sent to a networked reconstruction PC.

The scanner created an XML file with scan details (.xtekct) and 3142 projections in 16-bit greyscale TIFF format. Nikon CT Pro software was then used to calculate COR automatically and subsequently perform tomographic reconstruction through filtered back projection. Tomographic datasets were exported by this software as VGI and VOL files (VGStudio MAX format); these were imported into ImageJ (see Section 5.6.1.3), contrast adjusted and then converted into a series of 8-bit greyscale bitmap-format (.bmp) image files for reconstruction and visualization by the SPIERS software suite (Section 5.6.1.1). Voxel sizes ranging from 4 to 25 μm were achieved. Of the fossils scanned (around ~100 specimens to date), roughly a third have revealed enough novel morphology to justify scientific investigation.

3.2.13.2 Lab-Based Phase-Contrast X-Ray Microtomography: A Phoretic Mite in Amber

One form of three-dimensional preservation which CT can greatly aid is the study of fossils in amber. This is especially true of very small fossils, those in opaque amber or those obscured by other objects in the light path, hampering traditional microscopy. A recent example of lab-based high-resolution phase-contrast CT is Dunlop et al. (2012), which reports the reconstruction of a phoretic mite nymph hitching a ride on a spider's carapace (Figure 3.20). The composite fossil is hosted in Eocene (44–49 Ma) Baltic amber. The scans revealed both appendage details and a sucker plate, allowing description of one of the few convincing fossils of an astigmatid mite and providing a minimum age for the evolution of phoretic behaviour in the group's juveniles.

The scanned mite is 176 μm long, and scans were acquired with an Xradia MicroXCT system at the Manchester X-ray Imaging Facility, University of Manchester. Two scans each of 1200 projections were acquired with energies of 40 and 75 keV (the acceleration voltage and filament current are not reported). The scans were conducted at 10× and 20× optical magnifications providing voxel sizes of 1.7 and 0.87 μm. Furthermore, for each scan, propagation phase contrast was exploited to increase edge contrast. Source–object distance is reported in the further methodological details of Penney et al. (2011) as 250 mm and the detector–object distance as 150 mm. Tomographic reconstruction was conducted with Xradia TXMReconstructor software, and tomograms were generated as 16-bit TIFF files prior to visualization in Avizo 6.1.1 (Section 5.6.1).

3.2.13.3 Phase-Contrast Synchrotron Tomography: Conodonts

Goudemand et al. (2011) employed propagation-based phase-contrast X-ray microtomography in the study of three-dimensionally preserved, clustered oro-pharyngeal elements from conodonts. This was necessitated

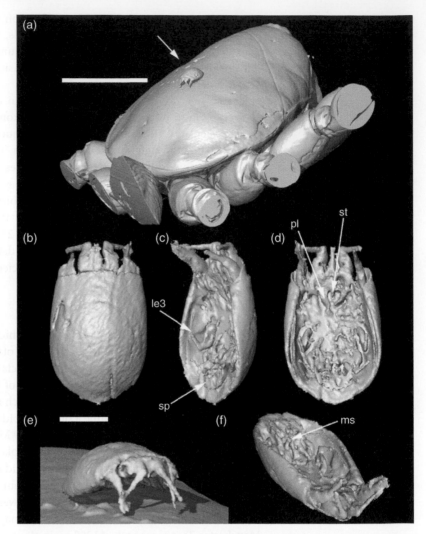

Figure 3.20 A phoretic mite found in amber on the dorsal carapace of a spider and reconstructed using phase-contrast nanotomography. (a) Mite in situ on the spider's carapace, indicated by arrow. (b–f) High-resolution scans of the mite. le 3- leg 3, ms = movable suckers, p1 = apodemes, sp = sucker plate, st = sternum. (a) Scale bars 500 μm, (b–f) scale bars 50 μm. *Source*: Dunlop et al. (2012, Fig. 1). Reproduced with permission of the Royal Society.

by a small sample size, thus need for high resolution and lack of attenuation contrast. The fossils of the Triassic conodont *Novispathodus* (Tiandong District, Guangxi Province, China) were found as fused clusters in which the sub-millimetre mouthparts retained their relative 3D positions and orientations (Figure 3.21). The scans suggest the presence of lingual cartilage similar in form to that seen in extant jawless fish, supporting the assumption conodonts are crown vertebrates.

Specimens were scanned at ID19 (ESRF) using a non-monochromatic 'pink beam' to increase flux and reduce acquisition time. This had a critical energy (that which bisects the emitted spectrum into parts of equal emitted power) of 17.68 keV and a narrow bandwidth obviating the need for a monochromator; this was delivered by an undulator (see Section 3.2.7.1). The detector comprised a 6 μm thick GGG ($Gd_3Ga_5O_{12}$) scintillator and a

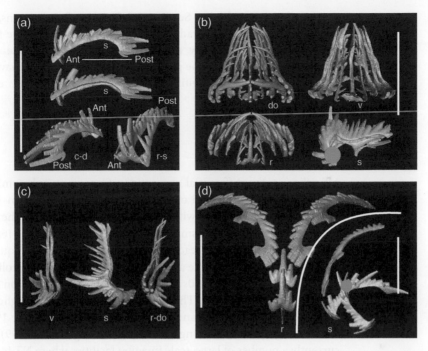

Figure 3.21 Conodont oro-pharyngeal elements reconstructed with the aid of phase-contrast synchrotron tomography. This figure shows the geometric correspondences between these elements. (a) S1 elements match the posterior process of the S0. (b) Closed arrangement of S elements. (c) Presumed growth position of the S2 (silver) as inferred by geometric correspondences with the S3 and S4 elements (gold). (d) Proposed movement of the S0 against the M elements (silver: start and end positions; gold: pinching position). Ant = anterior, post = posterior, c = caudal, d = dextral, do = dorsal, r = rostral, s = sinistral, v = ventral, blue circles = hypothetical cartilage. All scale bars 400 μm. *Source*: Goudemand et al. (2011, Fig. 3). Reproduced with permission of Nicolas Goudemand and the National Academy of Sciences.

FReLoN CCD camera. This provided voxel sizes of 0.23–0.46 μm, with a propagation distance of 10 mm. The authors do not report how the specimen was mounted. Phase retrieval was conducted with in-house algorithms (possibly those of Cloetens et al. 1999), including flat-field and dark-field corrections to reduce ring artefacts and sample movement correction. Tomograms were reconstructed using filtered back projection, and further ring artefact removal was conducted manually. The datasets were converted to 16-bit TIFF stacks for digital visualization with Amira (Section 5.6.1) and the in-house software FoRM-IT.

3.3 Neutron Tomography

NT is a non-destructive scanning technique for generating tomographic datasets of millimetre- to centimetre-sized samples at spatial resolutions down to several tens of micrometres. It works in an analogous way to X-ray computed tomography (see Section 3.2), but uses free neutrons, instead of X-rays, as the penetrating radiation. NT requires a large-scale facility as a neutron source, commonly a nuclear reactor or a particle accelerator (Vontobel et al. 2006). In the past decade, a few researchers have used this method to successfully digitize fossil material (e.g. Schwarz et al. 2005; Winkler 2006).

3.3.1   History

NT relies on the differential absorption of neutrons, subatomic particles that are found in the nuclei of all atoms apart from hydrogen-1. The first radiographic images made with neutrons were published in the late 1940s; these were produced using small accelerator sources (Peter 1946; Kallmann 1948). In the mid-1950s, reactor-based images, superior in quality due a higher neutron flux, became available (Thewlis 1956). The proliferation of research reactors in the 1960s drove a broader usage of neutron imaging in engineering and research. The development of modern, higher-intensity sources (see Section 3.3.2) and digital imaging systems in the 1990s facilitated tomographic work using neutrons (Schillinger et al. 1999). There are currently a number of large-scale imaging facilities where NT is performed (Strobl et al. 2009), but the technique is not yet as widely used as X-ray-based tomography.

3.3.2   Principles and Practicalities

Neutron imaging can be seen as a complementary alternative to X-ray imaging. Neutrons have no net charge; therefore, unlike X-rays, they do not interact with the electrons of atoms and instead directly probe atomic nuclei. Because nuclei are substantially smaller than atoms, matter is mostly empty space to neutrons. Neutron attenuation is caused by nuclear scattering or absorption. The size of the interaction varies haphazardly across the periodic table and is not correlated with density. This differentiates it from X-ray attenuation (see Section 3.2.3.2) and makes it appropriate for samples which are difficult to analyse with X-ray CT. In particular, neutrons are strongly attenuated by certain light elements, such as hydrogen, but readily penetrate many heavy elements, like lead. Neutrons are hence well suited for detecting hydrogen-containing materials within metals (Vlassenbroeck et al. 2007). In addition, neutrons generally penetrate matter much more easily than X-rays and so can be used to investigate larger samples.

In order to obtain an image using neutrons, a suitable source is required. Today, these are typically research reactors (e.g. the Forschungs-Neutronenquelle Heinz Maier-Leibnitz, or FRM II, reactor, Munich, Germany) in which neutrons are produced by nuclear fission or high-flux spallation sources at particle accelerators (e.g. the Swiss Spallation Neutron

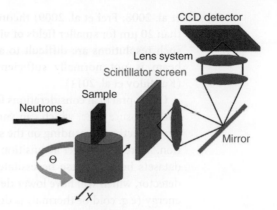

Figure 3.22 Typical NT set-up. *Source*: Strobl et al. (2009, Fig. 4). Reproduced with permission of the Institute of Physics Publishing.

Source, or SINQ, at the Paul Scherrer Institute, Switzerland), where the collision of high-energy protons with a target material causes neutrons to be emitted. The resulting *free neutrons* (i.e. neutrons that are not contained in an atomic nucleus) are slowed down by a moderator – for example, liquid deuterium or water which give, respectively, lower-energy 'cold' neutrons (~4 meV) or higher-energy 'thermal' neutrons (~25 meV) – according to the experiment being performed (neutron attenuation, and thus contrast between materials, is energy dependent). These neutrons are shaped into a low-divergence (i.e. narrow) beam using a collimator, and the extracted neutron beam is passed through the object of interest while it rotates 180° or 360° around a fixed axis (Figure 3.22). NT can then be performed in an analogous manner to CT (see Section 3.2.4.8); digital imaging detectors, frequently scintillator screens and CCD cameras, are used to acquire a large number of images (projections) of the object at different angles. These images record the extent to which the neutron beam is attenuated by the sample; filtered back projection (see Section 3.2.8.1) is used to mathematically reconstruct a tomographic dataset that maps the variation of neutron attenuation within the object.

The spatial resolution that can be attained using NT is governed by a number of different factors. Chief among these is neutron beam divergence, with higher beam divergence resulting in lower-resolution images. Beam divergence varies according to the configuration of the collimator and can be especially problematic for large objects, where the distance between the sample and the detector is usually substantial (e.g. tens of centimetres). The detector system is another important consideration. Larger detectors have higher detection efficiency (i.e. most neutrons passing through them will be detected), but will generate lower-resolution datasets because the specific point in the detector where the neutron was detected is unknown. Most modern systems have intrinsic limitations (e.g. thickness of the scintillator screen) that restrict them to resolutions of several tens of micrometres for fields of view of several tens of centimetres. Recent improvements of detector technologies (e.g. Siegmund et al. 2007; Tremsin et al. 2008; Frei et al. 2009) theoretically enable spatial resolutions of less than 20 μm for smaller fields of view (up to a few tens of millimetres), but such resolutions are difficult to achieve in practice because the neutron beam is not normally sufficiently intense to supply adequate signal (Kardjilov et al. 2011).

Other practical considerations for NT include the number of projections and exposure time, which can vary between seconds and several minutes per projection, depending on the sample properties and the desired resolution. Moreover, longer acquisition times are required for higher-resolution datasets because these necessitate the use of a smaller, better resolution detector, which will have lower detection efficiency. The choice of neutron energy (e.g. cold or thermal) is dependent on the composition and size of the studied object, and larger, poorly transmitting specimens will generally require higher beam energies. Finally, radioactive activation may be induced by neutron bombardment if the sample is rich in certain elements (e.g. cobalt or europium), necessitating internment of the specimen for several months or even years after the experiment. Such samples are therefore poorly suited to NT, and this could be a major drawback of the method in some fields.

3.3.3 Examples in Palaeontology

In recent years, a handful of workers have used NT to study fossil material, predominantly large vertebrates (see below), but occasionally organically preserved plants (e.g. Winkler 2006). Schwarz et al. (2005) were among the first to apply NT to fossils, using the SINQ spallation source at the Paul Scherrer Institute in Switzerland to image sauropod dinosaur remains from the Late Jurassic Morrison Formation of Wyoming, USA. They applied a thermal neutron flux of 3.6×10^6 neutrons/cm²/s and acquired 240–300 projections for each sample (acquisition times of 1.5–3 hours), generating tomographic datasets with resolutions of 272 μm. Unfortunately, they were unable to recover sharp contrast between the apatite fossil and the infilling siliciclastic matrix (due to insufficient neutron attenuation variation) (Figure 3.23a). Witzmann et al. (2010) used the CONRAD imaging facility at the BER-II research reactor at Helmholtz-Zentrum Berlin in Germany to investigate the bones of a range of basal tetrapods. Employing a cold neutron flux of approximately 10^7 neutrons/cm²/s, they were able to generate datasets with resolutions of about 200 μm per pixel and could therefore clearly visualize internal vascular canals within the studied fossils (Figure 3.23b). Comparable datasets (with similar spatial resolutions) have been obtained for other Mesozoic vertebrates using SINQ (Scheyer 2008) and CONRAD (Laaß et al. 2011), as well as research reactors in South Africa (SAFARI-1 reactor; Cisneros et al. 2010) and South Korea (HANARO reactor; Grellet-Tinner et al. 2011).

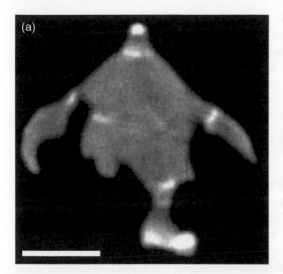 

Figure 3.23 (a) NT tomogram of a vertebra of an undetermined diplodocid sauropod. Scale bar is 20 mm. *Source*: Schwarz et al. (2005, Fig. 4). Reproduced with permission of the Society of Vertebrate Paleontology. (b) NT tomogram of the clavicle of the temnospondyl amphibian *Mastodonsaurus giganteus*. Scale bar is 30 mm. *Source*: Witzman et al. (2010, Fig. 4). Reproduced with permission of John Wiley and Sons.

3.3.4 Summary

NT is a relatively recent addition to the diverse array of tomographic techniques currently available to palaeontologists, having only become accessible in the last decade or so with the continued development of dedicated high-flux neutron sources throughout the world. Current facilities are capable of imaging large samples (e.g. up to about 0.3 m) at resolutions down to about 30–50 μm; however, palaeontological studies have typically attained resolutions on the order of a few hundred micrometres (e.g. Schwarz et al. 2005; Witzmann et al. 2010). NT relies on interior attenuation variations to generate an informative tomographic dataset, in the same way as X-ray CT. The absorption profile of neutrons renders NT particularly well suited to the study of large fossils hosted in dense, metal-rich rocks, as well as organically preserved specimens; such material can be difficult or impossible to image using X-ray CT. However, the best resolutions that can be achieved using NT are inferior to those of X-ray CT (see Schwarz et al. 2005 for a direct comparison), and NT acquisition times are typically longer. Neutron imaging facilities are also not as accessible as CT scanners, and the potentially hazardous levels of radioactivity that can be induced in certain samples could prohibit study of valuable specimens for months or years. We hence recommend NT primarily as a technique for material that has proved intractable to X-ray CT. Such fossils might include those surrounded by thick layers of dense materials or those that lack internal density contrast yet display variation in the distribution of neutron-attenuating elements (such as hydrogen). The development of new approaches including monochromatic neutron beams and phase-contrast tomography (Dubus et al. 2002) will enable higher-quality imaging of samples with limited neutron attenuation contrast, further expanding the palaeontological applications of NT.

3.4   Magnetic Resonance Imaging

MRI is a complex tomographic technique that uses a strong magnetic field to non-destructively map the nuclei of certain elements (usually hydrogen) within a sample. This information can be used to generate detailed images of the interior of the specimen under study; modern machines are capable of producing an isotropic dataset with a spatial resolution of less than 100 µm. MRI is well suited for imaging biological soft tissues, which contain abundant hydrogen atoms in the form of water molecules, and it has been an important clinical diagnostic tool for several decades (Keevil 2001). To date, however, very few studies have applied the technique to the study of fossilized specimens (e.g. Gingras et al. 2002; Clark et al. 2004; Mietchen et al. 2008).

3.4.1   History

MRI traces its history back to the late 1930s and 1940s, when the principle of **nuclear magnetic resonance (NMR)** – the physical phenomenon used in MRI – was first described (Rabi et al. 1938; Bloch et al. 1946; Purcell et al. 1946). Methods for producing images based on NMR were developed in the 1970s (Damadian 1971; Lauterbur 1973), and whole-body MRI scans were performed on humans soon after (Damadian et al. 1977; Mansfield et al. 1978). Subsequently, the use of MRI scanners became widespread in medicine, and there are currently tens of thousands of diagnostic systems located in hospitals worldwide. MRI is also extensively used in scientific research (there are probably several thousand pre-clinical scanners in the world), with applications in a broad range of fields, including biology, fluid mechanics and chemistry (Glover and Mansfield 2002; Tyszka et al. 2005; Carlson 2006; Ziegler et al. 2011).

3.4.2   Principles and Practicalities

Hydrogen is the element most commonly imaged using MRI because of its high magnetic sensitivity and natural abundance in biological systems. The most common form of hydrogen, the isotope ^{1}H, has a nucleus containing one proton and no neutrons; this unpaired proton gives the nucleus a net

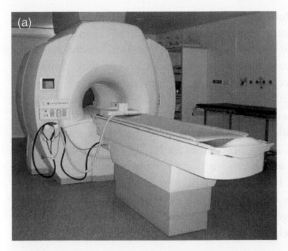
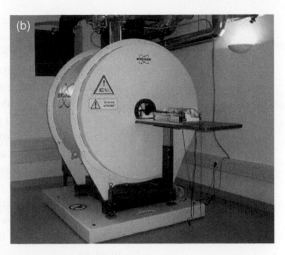

Figure 3.24 Magnetic resonance imaging (MRI) scanners. (a) Human scanner (3 T Philips Achieva) located in the Translational Research Imaging Center, Westfälische Wilhelms-Universität Münster. *Source*: http://campus.uni-muenster.de/3681.html. Reproduced with permission of Cornelius Faber. (b) Small-animal scanner (7 T Bruker Pharmascan) located in the Neuroscience Research Centre, Charité - Universitätsmedizin Berlin. *Source*: http://www.charite.de/nwfz/deutsch/animal-mrt/manual/index.html. Reproduced with permission of Susanne Mueller.

spin and thus a net magnetic moment. Standard MRI scanners consist of a magnet and a series of coils; during a scan, a static magnetic field aligns the magnetic moments of the hydrogen nuclei with the direction of the field, thus magnetizing the sample. Next, a radio-frequency (RF) pulse, with a frequency at or close to the resonance frequency of ^1H (which is proportional to the strength of the magnetic field), is applied to the sample, exciting the protons and perturbing the alignments of their magnetic fields. After the RF pulse has ended, the sample's magnetization relaxes to the original alignment of the static magnetic field, thereby generating a RF signal that is detected by the receiver coil surrounding the sample. If pulsed magnetic field gradients are also applied during the scan (so that magnetic field strength varies with location), spatial information will be encoded in this signal, and the three-dimensional distribution of excited protons in the sample can be recovered computationally. The end result is usually a three-dimensional tomographic dataset that maps hydrogen nuclei within the sample – two-dimensional tomograms can be produced for any desired plane by varying the application of the magnetic gradients and do not need to be registered prior to digital reconstruction.

Modern MRI scanners (Figure 3.24) are capable of imaging millimetre- to metre-sized specimens (governed by scanner type and coil size) at sub-millimetre resolutions. The resolution of MRI is dependent on a number of factors. Sample size is particularly important, with larger specimens requiring longer scanning times (up to tens of hours for solid samples with low hydrogen contents, such as many fossils) and larger coils, which limit the achievable spatial resolution. Secondly, the strength of the magnet can influence resolution, with higher field strengths (e.g. >3 T) increasing the resolution and SNR, but also the prevalence of artefacts – which may be minimized using software algorithms, although ideally they should be avoided during image acquisition by choosing appropriate scanning parameters (Smith and Nayak 2010; Ziegler et al. 2011). Resolutions typically range from about 10 to 100 μm in pre-clinical systems, but the development of nano-MRI scanners could make sub-micrometre resolutions attainable in the future (Mamin et al. 2013). Scan times are highly variable; datasets can take between minutes and hours to acquire, largely dependent on the sample's properties (e.g. hydrogen content) and the chosen imaging protocol. Note that materials exhibiting strong permanent magnetization (e.g. magnetic minerals) are generally unsuitable for MRI, as they will distort the homogenic magnetic field inside the scanner that is integral to successful imaging.

3.4.3 Examples in Palaeontology

MRI has been applied only rarely in palaeontology. McJury et al. (1994) used a clinical MRI scanner to image centimetre-sized dinosaur eggs from the Upper Cretaceous of Hunan Province, China, taking advantage of water held in the egg shells. The duration of this scan was over 100 minutes, and the resulting tomograms had a fairly coarse resolution of about 0.01×0.2 m (slice thickness not specified). Gingras et al. (2002) investigated trace fossil burrows in sandstone from the Lower Cretaceous of Alberta, Canada, which contained residual water in pore spaces and so were amenable to MRI. They were able to image a $0.07 \times 0.04 \times 0.03$ m block of sandstone in 55 minutes, obtaining images with resolutions of about 470 μm. Finally, Mietchen et al. (2005, 2008) applied MRI to a range of different fossil taxa, including belemnites, a crinoid, a partial skeleton of a dolphin and silicified plants (Figure 3.25). Using a pre-clinical scanner (with a high magnetic field strength), they visualized the anatomy of specimens based on the presence of intracrystalline water. They were able to obtain tomograms with a far higher resolution (80 μm) than previous palaeontological studies, but due to the low signal strengths of the studied samples, acquisition times were consequently much longer (18–93 hours).

Alternatively, or in addition, the hydrogen content can be artificially increased in fossil samples through the use of contrast agents. For example, Steiger (2001) immersed a fossil radiolarian in silicon oil to increase ^1H, thereby enhancing the resonance signal. Clark et al. (2004) were able to achieve the same effect by filling a mouldic fossil dicynodont from the Permian of Scotland with water and imaging the cavity left by the original skeleton using a clinical MRI scanner. They acquired a tomographic dataset in about four hours with a resolution of 1 mm.

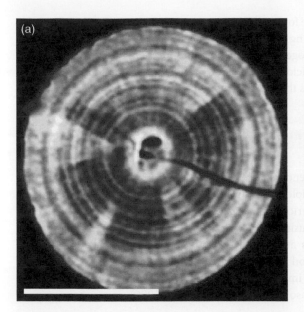
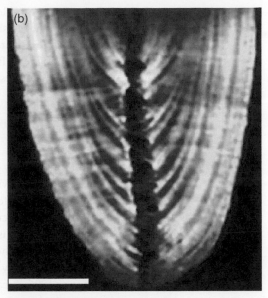

Figure 3.25 (a, b) MRI tomograms of the belemnite *Belemnopsis* sp. *Source*: Mietchen et al. (2008, Fig. 2). Scale bars are 5 mm. Reproduced from the original images, which are licensed under a Creative Commons Attribution 3.0 License (http://creativecommons.org/licenses/by/3.0/).

3.4.4 Summary

Although MRI has traditionally been thought to be wholly inappropriate for palaeontological samples, several studies in the last few years have shown that this modality can resolve sub-millimetre morphological structures in fossils. The technique works best when hydrogen content is high, and so fossils with residual water (e.g. McJury et al. 1994; Gingras et al. 2002) or those immersed in solutions containing ^{1}H (e.g. Steiger 2001; Clark et al. 2004) are better suited for MRI than other samples. Mummified and frozen specimens, in particular, should contain relatively large amounts of water and hence could be amenable to MRI (Mietchen et al. 2008). Alternatively, MRI could be used to image elements other than hydrogen, such as carbon (^{13}C) or phosphorous (^{31}P), which might be more common in certain fossils (although perhaps not sufficiently abundant to produce a strong resonance signal). Unfortunately, MRI systems are often expensive to use and might not be widely accessible to palaeontologists, limiting their utility for scientific research. Furthermore, the contrast and resolution of MRI images acquired for fossils do not currently compete with the state of the art for other tomographic techniques (such as X-ray computed tomography), making it a sub-optimal modality for the three-dimensional visualization of fossil anatomy. However, the NMR signal that is detected using MRI potentially provides information on three-dimensional chemical composition that might not be obtainable using other techniques. Therefore, MRI could prove extremely valuable for elucidating, for example, the taphonomic processes responsible for fossilization. In summary, many fossils will remain difficult to image using MRI, as they typically produce a weak NMR signal due to their low hydrogen content; future development of systems capable of stronger magnetic field gradients and pulses, as well as enhanced SNR, might help to address this fundamental impediment (Carlson 2006).

3.5 Optical Tomography: Serial Focusing

Optical tomography encompasses a range of techniques that involve shining light (visible, ultraviolet or near infrared) through a sample to acquire tomograms of its interior structures. The most useful approach for palaeontologists is serial focusing (also known as optical sectioning), which uses a microscope to focus at successive depths in a translucent sample. This serial-focusing method is often performed with confocal microscopy, which enables high-quality imaging of fossils at sub-micrometre resolutions (e.g. O'Connor 1996; Ascaso et al. 2003; Schopf et al. 2006).

3.5.1 *History*

Confocal microscopy works by illuminating and detecting single focal planes within a sample through the elimination of out-of-focus light, thereby reducing blur compared to conventional light or fluorescence microscopy. The design for the first confocal microscope was patented over half a century ago (Minsky 1961; described in Minsky 1988). Subsequently, a number of instruments based on this concept were developed, including the tandem-scanning confocal microscope (Petráň et al. 1968), which uses a rapidly spinning disk with a series of holes to image the specimen; the slit-scanning confocal microscope (Svishchev 1969), which employs a slitlike light source for imaging; and the laser-scanning confocal microscope (Brakenhoff et al. 1979), which uses a focused laser beam to scan the object (the history of the laser is outlined in Section 4.2.1). **Confocal laser-scanning microscopy** (CLSM) became increasingly popular in the late 1980s, when a number of commercial microscopes were made available (White and Amos 1987), and it is now widely used in the life and materials sciences (Amos and White 2003; Halbhuber and König 2003; Hovis and Heuer 2010).

3.5.2 *Principles and Practicalities*

3.5.2.1 Confocal Laser-Scanning Microscopy

CLSM is the most common type of confocal microscopy used in palaeontological research. Depending on the properties of the sample being imaged, CLSM can be performed in fluorescence and/or reflectance modes, and using different wavelengths of laser light. In both cases, the microscope uses

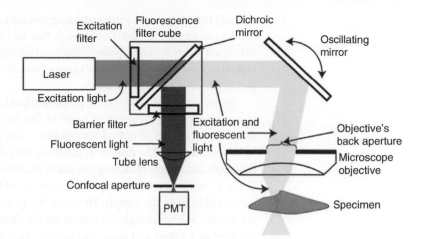

Figure 3.26 Schematic diagram of a confocal laser-scanning microscope. *Source*: Conchello and Litchtman (2005, Fig. 1). Reproduced with permission of Nature Publishing Group.

a point laser source (see Section 4.2.2 for a description of laser technologies), which is focused onto the sample by an objective lens, to intensely illuminate a small, sub-micrometre spot on the plane of interest. The spot is then scanned over the sample (typically using a pair of mirrors) and light fluorescently emitted and/or reflected from the specimen (Figure 3.26). Fluorescence emission is characterized by a loss of energy as the absorbed light shifts the fluorescent molecules into a higher energy state. This energy is lost through the emission of a fluorescent photon at a longer wavelength than the excitation wavelength (Stokes fluorescence, usually described as Stokes shift). The emitted light is re-collected by the objective lens and returned through the optical pathway. When it meets the dichroic mirror (Figure 3.26), only light at a wavelength longer than that of the laser is transmitted to the detector apparatus. This mechanism prevents the detector (typically a photomultiplier tube) from being overwhelmed by signal at the same wavelength as the laser and also separates reflected laser light from fluorescent signal emission. The use of a pinhole inserted into the light path, in front of the detector, at a point conjugate to the **focal plane**, blocks the passage of most light from out-of-focus planes. In reflection-mode confocal microscopy, the procedure is similar, but the signal intensity from reflection between the laser and the sample is mapped (instead of fluorescence), and the dichroic mirror is omitted from the optical path. Again, the pinhole serves to exclude defocused signal from the detector.

Since only a single spot of light is produced at any one time, the illuminated spot is represented by a single pixel in an image electronically generated by the detector; the laser beam is scanned across the focal plane in a series of lines to image the entire tomogram. Scan times of less than 1 μs per pixel are feasible, meaning that a 512 × 512 pixel image can be acquired in less than 1 second (Conchello and Lichtman 2005); however, if the SNR is low, it may be necessary to obtain multiple exposures for each pixel, and this can substantially slow down image acquisition. Video-rate (30 frames per second) imaging can be achieved using specially designed microscopes (e.g. Sanderson and Parker 2003), although this sort of ultra-rapid acquisition is of limited utility for static palaeontological specimens. When using standard objective lenses, the field of view ranges from about 10 μm to a few millimetres depending on magnification.

This two-dimensional imaging procedure can be repeated for multiple focal planes by adjusting the depth of focus in fine increments, producing a three-dimensional dataset of optical tomograms – this is termed serial focusing or optical sectioning. The process is generally non-destructive, but fluorescence depletion and sample bleaching can occur in certain samples. Image resolution in the x- and y-axes is related to the resolution of the lens and the wavelength used to illuminate the sample. However, the practical resolution in the z-axis (i.e. depth) will be strongly dependent on the degree of scatter experienced by the laser as it enters and leaves the sample. This scatter reduces the SNR and increases with depth into the specimen, depending on the optical clarity of the sample. Additionally, the choice of objective lens can affect depth penetration: lower magnification lenses are capable of greater working distances, but at reduced resolutions. Artefacts (e.g. blurring or darkening in tomograms) may arise if the light path is partially blocked by extraneous objects in the sample. Under optimal conditions, blur-free images with horizontal resolutions of less than 0.2 μm and section thicknesses of less than 0.4 μm can be obtained with high-magnification lenses. Moreover, imaging depths of up to about 200–300 μm can be achieved for relatively transparent specimens. After tomograms have been acquired for a volume of material, digital reconstruction is performed as standard (see Chapter 5); registration of images is not normally required, as CLSM set-ups can adjust focal depth without altering sample position.

Confocal microscopes use a sequence of detectors and mirrors to pass light of different wavelengths to the detectors and thus separate out different emission wavelengths. This allows them to collect signals from parts of the sample which have different fluorescent responses. This is important for biological research where the signals detected are usually the result of tagging regions of the sample with fluorescently labelled antibodies to mark different targets. In the examination of palaeontological samples, however, the signal detected is usually a result of the autofluorescent (naturally fluorescing) properties of the sample. It is useful to know which matching set of excitation wavelengths and emission spectral 'windows' gives the best result on a given sample. Many confocal microscopes can carry out a spectrographic analysis of the sample to generate a graph indicating the optimum setting for the laser/detector pairing. In some cases, this process can result in live spectrographic images being generated, and fluorescence in undesirable wavelengths can be removed (spectral unmixing) to give images which highlight the features of most interest.

3.5.2.2 Confocal Raman Imagery

Confocal microscopy can be combined with **Raman spectroscopy** in order to obtain three-dimensional information about the chemical structure of a sample. This is termed confocal Raman imagery. Like CLSM, the technique uses laser light to illuminate a small spot on the sample, exciting molecular vibrations through the inelastic (Raman) scattering of photons. The detected signal is represented as a Raman spectrum, which reflects the molecular composition of the studied sample. For two-dimensional imagery, a complete spectrum is collected for each pixel of the image, producing an image that consists of hundreds or thousands of Raman spectra. When this process is repeated for multiple focal planes, through serial focusing, a three-dimensional volumetric dataset can be built up. The spatial resolution of confocal Raman imagery (e.g. ~1 µm) is lower than that of CLSM, and the application of the method is also much more time-consuming (it can take over one hour to obtain all of the point spectra that contribute to a single two-dimensional slice). However, the method does provide direct information on sample composition that cannot be obtained using CLSM.

3.5.3   Examples in Palaeontology

Since the early 1990s, serial focusing with CLSM has been (infrequently) used to study palaeontological material. Much work has focused on acid-macerated fossil palynomorphs, including dinoflagellate cysts (Feist-Burkhardt and Pröss 1999), acritarchs (Talyzina 1998; Martí Mus and Moczydłowska 2000), spores (Scott and Hemsley 1991) and pollen (Nix and Feist-Burkhardt 2003; Hochuli and Feist-Burkhardt 2004). Such fossils are well suited for CLSM because of their small size (<1 mm), thin, light-transmitting outer walls and high content of autofluorescent organic material. A few workers have used CLSM to analyse microfossils preserved in small pieces of amber, notably fungi from the Lower Cretaceous of Álava, Spain (Ascaso et al. 2003, 2005; Speranza et al. 2010). These studies employed a Zeiss LSM 310 laser-scanning confocal microscope equipped with a 63× oil-immersion objective. Argon and helium–neon lasers, with excitation wavelengths of 488 nm and 543 nm, respectively, were used to scan polished specimens mounted in slides; because the amber is translucent, it was possible to excite autofluorescence in the included fossils to obtain sharp, high-resolution optical sections (in stacks of 20–30 images). Compton et al. (2010) detail another application of CLSM to amber, studying fig wasps hosted in Miocene Dominican amber using a Leica TCS SP1 confocal microscope and an argon laser with an excitation wavelength of 488 nm.

The most extensive use of CLSM in palaeontology was by Schopf and colleagues, who imaged thin sections containing rock-embedded microfossils from the Cambrian of China and the Precambrian of Australia, Canada, India and Kazakhstan (e.g. Schopf et al. 2006, 2008; Chen et al. 2007; Schopf and Kudryavtsev 2009, 2011). Prior to imaging, they covered samples with a thin veneer of Cargille immersion oil (type FF, fluorescence free). An Olympus FluoView 200 laser-scanning confocal

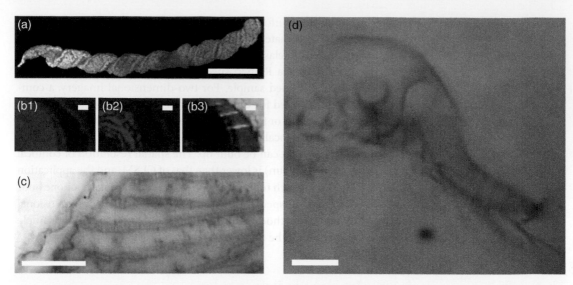

Figure 3.27 Optical tomograms. (a) Confocal laser-scanning image of the oscillatoriacean cyanobacterium *Heliconema funiculum* from the Precambrian Bitter Springs Formation of Australia. Scale bar is 10 μm. *Source*: Schopf & Kudryavtsev (2009, Fig. 3). Reproduced with permission of Elsevier. (b) Confocal Raman images of a ctenophore embryo from the Cambrian Kuanchuanpu Formation of China showing the distribution of (b1) kerogen in blue, (b2) apatite in red and (b3) calcite in green. Scale bars are 10 μm. *Source*: Schopf & Kudryavtsev (2009, Fig. 2). Reproduced with permission of Elsevier. (c) Serial-focusing image (acquired with light microscopy) of the book lungs of the trigonotarbid arachnid *Palaeocharinus* sp. from the Devonian Rhynie chert of Scotland. Scale bar is 20 μm. *Source*: Kamenz et al. (2008, Fig. 1). Reproduced with permission of Royal Society Publishing. (d) Serial-focusing image (acquired with bright-field microscopy) of an undescribed branchiopod from the Devonian Windyfield chert of Scotland. Scale bar is 100 μm. *Source*: Haug et al. (2009, Fig. 3). Reproduced with permission of John Wiley and Sons.

microscope with a 60× oil-immersion objective was then used to study the specimens. An argon laser with an excitation wavelength of 488 nm and a helium–neon laser with an excitation wavelength of 633 nm were utilized to image autofluorescence (filters removed wavelengths <510 nm for the argon laser and <660 nm for the helium–neon laser). In this manner, high-resolution, three-dimensional anatomy was visualized for diverse taxa such as ctenophores and cyanobacteria (Figure 3.27a). Furthermore, they also studied many of the same samples with confocal Raman imagery to map their chemical structure. A Horiba T64000 Raman system was used to acquire Raman spectra of kerogen, apatite and calcite (Figure 3.27b).

3.5.4 Other Approaches

In addition to confocal microscopy, there are a number of other techniques that use optical imaging to non-destructively extract three-dimensional volumetric data from samples. Two of the most common approaches, optical coherence tomography and optical projection tomography, are better suited for biological tissues than mineralized samples; however, several less precise techniques have been successfully used for serial focusing of palaeontological material. For example, conventional light microscopes can acquire optical sections over a range of depths so long as a sufficiently high numerical aperture is used to restrict the depth of field (Agard 1984). Here, the focus is typically adjusted manually, in small, sub-micrometre steps, with photographs taken every step to produce many closely spaced sections – these optical tomograms may need to be digitally registered after acquisition to compensate for small horizontal displacements caused by altering the focus of the microscope.

Kamenz et al. (2008) used a Zeiss Axioplan 2 light microscope with a 100× oil-immersion lens to image the three-dimensional microanatomy of trigonotarbid book lungs from the Early Devonian Rhynie chert of Scotland (Figure 3.27c). They acquired 600 images at 0.1 μm intervals through a thin section of translucent chert, giving a total stack depth of 60 μm. Haug et al. (2009) studied fossil arthropods from the Early Devonian Windyfield chert of Scotland and the Upper Cambrian 'Orsten' fauna of Sweden using bright-field and dark-field microscopy, respectively (Figure 3.27d). They viewed samples with a Zeiss Axio Scope microscope and photographed sections with an Olympus E-20P digital camera. These specimens were optically sectioned at 0.9 μm intervals, yielding around 200 images per fossil.

3.5.5 Summary

Serial focusing is an optical tomographic technique for studying light-transmitting microfossils which are either isolated (e.g. acid-macerated palynomorphs) or hosted in a translucent matrix (e.g. chert or amber). When performed using CLSM, sharp images with sub-micrometre resolutions can be rapidly generated, revealing three-dimensional microanatomy (mostly) non-destructively. CLSM can be used to detect reflective or fluorescent light; reflectance imaging has not generally proven useful for fossil samples because high internal reflection often obscures the region of interest (O'Connor 1996), but fluorescence imaging is well suited for organically preserved fossils, which autofluoresce when excited by the appropriate wavelength of laser light. This can be combined with confocal Raman imagery to obtain quantitative information about the chemical composition of fossils (Schopf and Kudryavtsev 2011). Other potential applications include probing for biogenic growth markers in biomineralized fossils and examining seasonal algal deposits in caves (which could also leave a fluorescent trace). Conventional light microscopy is an alternative method for achieving serial focusing at comparable resolutions; it has a higher depth of field (reducing image sharpness) and more limited depth penetration than CLSM, but the wider availability of light microscopes makes this an accessible approach. All serial-focusing methods are very restricted in their utility, however. Most fossil samples are opaque or transmit light poorly, making them unsuitable for serial focusing, and even for 'clean' translucent materials, imaging cannot be carried out at depths below about 200–300 μm. If the light path is blocked by extraneous material, artefacts will be prevalent in tomograms, and this could be a common problem for matrix-hosted fossils (e.g. Rhynie chert). Serial focusing should therefore be taken as a niche technique for investigating suitably preserved fossils (especially palynomorphs and other organic-walled microfossils) at high resolutions without causing damage to the specimen, and will be particularly useful for sub-millimetre-sized fossils in thin sections or acid-macerated preparations that do not show sufficient X-ray absorption contrast for computed tomography.

References

Abel, R.L., Laurini, C. & Richter, M. (2012) A palaeobiologist's guide to "virtual" micro-CT preparation. *Palaeontologia Electronica*. **15 (2)**, 6T, 17p.

Agard, D.A. (1984) Optical sectioning microscopy: cellular architecture in three dimensions. *Annual Review of Biophysics and Bioengineering*, **13**, 191–219.

Als-Nielsen, J. & McMorrow, D. (2011) *Elements of Modern X-Ray Physics*. Wiley-Blackwell, Malden.

Amos, W.B. & White, J.G. (2003) How the confocal laser scanning microscope entered biological research. *Biology of the Cell*, **95 (6)**, 335–342.

Ando, M., Chen, J. & Hyodo, K. (2000) Nondestructive visual search for fossils in rock using X-ray interferometry imaging. *Japanese journal of applied physics. Pt. 2, Letters*. **39 (10)**, 1009–1011.

Arnold, J.R., Testa, J.P., Friedman, P.J., et al. (1983) Computed tomographic analysis of meteorite inclusions. *Science*, **219**, 383–384.

Ascaso, C., Wierzchos, J., Corral, J.C., et al. (2003) New applications of light and electron microscopic techniques for the study of microbiological inclusions in amber. *Journal of Paleontology*, **77 (6)**, 1182–1192.

Ascaso, C., Wierzchos, J., Speranza, M., et al. (2005) Fossil protists and fungi in amber and rock substrates. *Micropaleontology*, **51 (1)**, 59–72.

Barrett, J.F. & Keat, N. (2004) Artifacts in CT: recognition and avoidance. *Radiographics*, **24**, 1679–1691.

Betz, O., Wegst, U. & Weide, D. (2007) Imaging applications of synchrotron X-ray phase-contrast microtomography in biological morphology and biomaterials science. I. General aspects of the technique. *Journal of Microscopy*, **227**, 51–71.

Beuck, L., Wisshak, M., Munnecke, A., et al. (2008) A giant boring in a Silurian stromatoporoid analysed by computer [sic] tomography. *Acta Palaeontologica Polonica*, **53 (1)**, 149–160.

Bloch, F., Hansen, W.W. & Packard, M. (1946) The nuclear induction experiment. *Physical Review*, **70 (7–8)**, 474–485.

Boas, F.E. & Fleischmann, D. (2012) CT artifacts: causes and reduction techniques. *Imaging in Medicine*, **4 (2)**, 229–240.

Brabant, L., Vlassenbroeck, J., De Witte, Y., et al. (2011) Three-dimensional analysis of high-resolution X-ray computed tomography data with Morpho+. *Microscopy and Microanalysis*, **17 (2)**, 252–263.

Brakenhoff, G.J., Blom, P. & Barends, P. (1979) Confocal scanning light microscopy with high aperture immersion lenses. *Journal of Microscopy*, **117 (2)**, 219–232.

Bushberg, J.T., Seibert, J.A., Leidholdt, E.M., et al. (2011) *The Essential Physics of Medical Imaging*. Lippincott Williams & Wilkins, Philadelphia.

Buzug, T.M. (2008) *Computed Tomography: From Photon Statistics to Modern Cone-Beam CT*. Springer, New York.

Carlson, W.D., Rowe, T.B., Ketcham, R.A., et al. (2003) Applications of high-resolution X-ray computed tomography in petrology, meteoritics and palaeontology. In: Mees, F., Swennen, R., Geet, M.V., et al. (eds), *Applications of X-Ray Computed Tomography in the Geosciences*, pp. 7–22. Geological Society, London.

Carlson, W.D. (2006) Three-dimensional imaging of earth and planetary materials. *Earth and Planetary Science Letters*, **249 (3–4)**, 133–147.

Chaimanee, Y., Jolly, D., Benammi, M., et al. (2003) A middle Miocene hominoid from Thailand and orangutan origins. *Nature*, **422**, 61–65.

Chen, J.-Y., Schopf, J.W., Bottjer, D.J., et al. (2007) Raman spectra of a lower Cambrian ctenophore embryo from southwestern Shaanxi, China. *Proceedings of the National Academy of Sciences of the United States of America*, **104 (15)**, 6289–6292.

Cisneros, J.C., Cabral, U.G., de Beer, F., et al. (2010) Spondarthritis in the Triassic. *PLoS ONE*, **5 (10)**, e13425.

Cifelli, R.L., Rowe, T.B., Luckett, W.P., et al. (1996) Fossil evidence for the origin of the marsupial pattern of tooth replacement. *Nature*, **379**, 715–717.

Clark, N.D.L., Adams, C., Lawton, T., et al. (2004) The Elgin marvel: using magnetic resonance imaging to look at a mouldic fossil from the Permian of Elgin, Scotland, UK. *Magnetic Resonance Imaging*, **22 (2)**, 269–273.

Cloetens, P., Barrett, R., Baruchel, J., et al. (1996) Phase objects in synchrotron radiation hard X-ray imaging. *Journal of Physics D: Applied Physics*, **29 (1)**, 133–146.

Cloetens, P., Ludwig, W., Baruchel, J., et al. (1999) Holotomography: quantitative phase tomography with micrometer resolution using hard synchrotron radiation X-rays. *Applied Physics Letters*, **75 (19)**, 2912–2914.

Compton, S.G., Ball, A.D., Collinson, M.E., et al. (2010) Ancient fig wasps indicate at least 34 Myr of stasis in their mutualism with fig trees. *Biology Letters*, **6 (6)**, 838–842.

Conchello, J.-A. & Lichtman, J.W. (2005) Optical sectioning microscopy. *Nature Methods*, **2 (12)**, 920–931.

Conroy, G.C. & Vannier, M.W. (1984) Noninvasive three-dimensional computer imaging of matrix-filled fossil skulls by high-resolution computed tomography. *Science*, **226**, 456–458.

Cooper, D.M.L., Chapman, L.D., Carter, Y., et al. (2012) Three dimensional mapping of strontium in bone by dual energy K-edge subtraction imaging. *Physics in Medicine and Biology*, **57 (18)**, 5777–5786.

Damadian, R. (1971) Tumor detection by nuclear magnetic resonance. *Science*, **171 (3976)**, 1151–1153.

Damadian, R., Goldsmith, M. & Minkoff, L. (1977) NMR in cancer: XVI. FONAR image of the live human body. *Physiological Chemistry and Physics*, **9 (1)**, 97–100.

Davis, G.R. & Elliott, J.C. (2006) Artefacts in X-ray microtomography of materials. *Materials Science and Technology*, **22 (9)**, 1011–1018.

Davis, G.R. & Wong, F.S. (1996) X-ray microtomography of bones and teeth. *Physiological Measurement*, **17 (3)**, 121–46.

Denison, C., Carlson, W.D. & Ketcham, R.A. (1997) Three-dimensional quantitative textural analysis of metamorphic rocks using high-resolution computed X-ray tomography: part I. Methods and techniques. *Journal of Metamorphic Geology*, **15 (1)**, 29–44.

DeVore, M.L. & Kenrick, P. (2006) Utility of high resolution X-ray computed tomography (HRXCT) for paleobotanical studies: an example using London Clay fruits and seeds. *American Journal of Botany*, **93 (12)**, 1848–1851.

Dierick, M., Cnudde, V., Masschaele, B., et al. (2007) Micro-CT of fossils preserved in amber. *Nuclear Instruments and Methods in Physics Research Section A: Accelerators, Spectrometers, Detectors and Associated Equipment*, **580**, 641–643.

Dominguez, P., Jacobson, A.G. & Jefferies, R.P.S. (2002) Paired gill slits in a fossil with a calcite skeleton. *Nature*, **417**, 841–844.

Donoghue, P.C.J., Bengtson, S., Dong, X., et al. (2006) Synchrotron X-ray tomographic microscopy of fossil embryos. *Nature*, **442**, 680–683.

Dubus, F., Bonse, U., Biermann, T., et al. (2002) Tomography using monochromatic thermal neutrons with attenuation and phase contrast. In: Bonse, U. (ed), *Developments in X-Ray Tomography III. Proceedings of SPIE*, **4503**, 359–370.

Dunlop, J.A., Wirth, S., Penney, D., et al. (2012) A minute fossil phoretic mite recovered by phase-contrast X-ray computed tomography. *Biology Letters*, **8 (3)**, 457–460.

Elder, F., Gurewitsch, A., Langmuir, R., et al. (1947) Radiation from electrons in a synchrotron. *Physical Review*, **71 (11)**, 829–830.

Elliott, J.C. & Dover, S.D. (1982) X-ray microtomography. *Journal of Microscopy*, **126 (2)**, 211–213.

Elliott, J.C., Dowker, S.E. & Knight, R.D. (1981) Scanning X-ray microradiography of a section of a carious lesion in dental enamel. *Journal of Microscopy*, **123 (1)**, 89–92.

Feist-Burkhardt, S. & Pross, J. (1999) Morphological analysis and description of Middle Jurassic dinoflagellate cyst marker species using confocal laser scanning microscopy, digital optical microscopy, and conventional light microscopy. *Bulletin du Centre de Recherches Elf Exploration Production*, **22 (1)**, 103–145.

Feldkamp, L.A., Goldstein, S.A., Parfitt, A.M., et al. (1989) The direct examination of three-dimensional bone architecture in vitro by computed tomography. *Journal of Bone and Mineral Research*, **4 (1)**, 3–11.

Flannery, B. & Deckman, H. (1987) Three-dimensional X-ray microtomography. *Science*, **237**, 1439–1444.

Frei, G., Lehmann, E.H., Mannes, D., et al. (2009) The neutron micro-tomography setup at PSI and its use for research purposes. *Nuclear Instruments and Methods in Physics Research A*, **605 (1–2)**, 111–114.

Friedland, G.W. & Thurber, B.D. (1996) The birth of CT. *American Journal of Roentgenology*, **167**, 1365–1370.

Friis, E.M., Crane, P.R., Pedersen, K.R., et al. (2007) Phase-contrast X-ray microtomography links Cretaceous seeds with Gnetales and Bennettitales. *Nature*, **450**, 549–52.

Fuchs, T.O.J., Kachelriess, M. & Kalender, W. (2003) Fast *volume scanning* approaches by X-ray-computed tomography. *Proceedings of the IEEE*, **91 (10)**, 1492–1502.

Garwood, R.J. & Dunlop, J.A. (2010) Fossils explained 58 – Trigonotarbids. *Geology Today*, **26 (1)**, 34–37.

Garwood, R.J. & Dunlop, J.A. (2011) Morphology and systematics of Anthracomartidae (Arachnida: Trigonotarbida) *Palaeontology*, **54 (1)**, 145–161.

Garwood, R.J., Dunlop, J.A., Giribet, G., et al. (2011) Anatomically modern Carboniferous harvestmen demonstrate early cladogenesis and stasis in Opiliones. *Nature Communications*, **2**, 444.

Garwood, R.J., Dunlop, J.A. & Sutton, M.D. (2009) High-fidelity X-ray micro-tomography reconstruction of siderite-hosted Carboniferous arachnids. *Biology Letters*, **5**, 841–844.

Garwood, R.J., Ross, A., Sotty, D., et al. (2012) Tomographic reconstruction of neopterous carboniferous insect nymphs. *PLoS ONE*, **7 (9)**, e45779.

Garwood, R.J. & Sutton, M.D. (2010) X-ray micro-tomography of Carboniferous stem-Dictyoptera: new insights into early insects. *Biology Letters*, **6**, 699–702.

Gingras, M.K., MacMillan, B., Balcom, B.J., et al. (2002) Using magnetic resonance imaging and petrographic techniques to understand the textural attributes and porosity distribution in *Macaronichnus*-burrowed sandstone. *Journal of Sedimentary Research*, **72 (4)**, 552–558.

Glover, P. & Mansfield, P. (2002) Limits to magnetic resonance microscopy. *Reports on Progress in Physics*, **65 (10)**, 1489–1511.

Görög, Á., Szinger, B., Tóth, E., et al. (2012) Methodology of the micro-computer [sic] tomography on foraminifera. *Palaeontologia Electronica*, **15 (1)**, 3T, 15p.

Goudemand, N., Orchard, M.J., Urdy, S., et al. (2011) Synchrotron-aided reconstruction of the conodont feeding apparatus and implications for the mouth of the first vertebrates. *Proceedings of the National Academy of Sciences*, **108 (21)**, 8720–8724.

Grellet-Tinner, G., Sim, C.M., Kim, D.H., et al. (2011) Description of the first lithostrotian titanosaur embryo *in ovo* with neutron characterization and implications for lithostrotian Aptain migration and dispersion. *Gondwana Research*, **20 (2–3)**, 621–629.

Grün, R., Schwarcz, H.P. & Zymela, S. (1987) Electron spin resonance dating of tooth enamel. *Canadian Journal of Earth Sciences*, **24 (5)**, 1022–1037.

Grün, R. & Stringer, C. (2007) Electron spin resonance dating and the evolution of modern humans. *Archaeometry*, **33 (2)**, 153–199.

Hagadorn, J.W., Xiao, S., Donoghue, P.C.J., et al. (2006) Cellular and subcellular structure of neoproterozoic animal embryos. *Science*, **314**, 291–294.

Halbhuber, K.-J. & König, K. (2003) Modern laser scanning microscopy in biology, biotechnology and medicine. *Annals of Anatomy*, **185 (1)**, 1–20.

Hall, C., Barnes, P. & Cockcroft, J. (1998) Synchrotron energy-dispersive X-ray diffraction tomography. *Nuclear Instruments and Methods in Physics Research Section B: Beam Interactions with Materials and Atoms*, **140**, 253–257.

Haubitz, B., Prokop, M. & Dohring, W. (1988) Computed tomography of Archaeopteryx. *Paleobiology*, **14 (2)**, 206–213.

Haug, J.T., Haug, C., Maas, A., et al. (2009) Simple 3D images from fossil and recent micromaterial using light microscopy. *Journal of Microscopy*, **233 (1)**, 93–101.

Helfen, L., Myagotin, A., Pernot, P., et al. (2006) Investigation of hybrid pixel detector arrays by synchrotron-radiation imaging. *Nuclear Instruments and Methods in Physics Research Section A: Accelerators, Spectrometers, Detectors and Associated Equipment*, **563**, 163–166.

Herzog, H. (2002) In vivo functional imaging with SPECT and PET. *Radiochimica Acta*, **214**, 203–214.

Hochuli, P.A. & Feist-Burkhardt, S. (2004) A boreal early cradle of angiosperms? Angiosperm-like pollen from the Middle Triassic of the Barents Sea (Norway). *Journal of Micropalaeontology*, **23 (2)**, 97–104.

Hounsfield, G.N. (1973) Computerized transverse axial scanning (tomography): part I. Description of system. *British Journal of Radiology*, **46**, 1016–1022.

Houssaye, A., Xu, F. & Helfen, L. (2011) Three-dimensional pelvis and limb anatomy of the Cenomanian hind-limbed snake Eupodophis descouensi (Squamata, Ophidia) revealed by synchrotron-radiation. *Journal of Vertebrate Palaeontology*, **31 (1)**, 2–7.

Hovis, D.B. & Heuer, A.H. (2010) The use of laser scanning confocal microscopy (LSCM) in materials science. *Journal of Microscopy*, **240 (3)**, 173–180.

Hsieh, J. (2003) *Computed Tomography: Principles, Design, Artifacts, and Recent Advances*. Wiley Inter-Science/SPIE Press, Bellingham.

Hsieh, J., Nett, B., Yu, Z., et al. (2013) Recent advances in CT image reconstruction. *Current Radiology Reports*, **1 (1)**, 39–51.

Jacques, S.D.M., Egan, C.K., Wilson, M.D., et al. (2013) A laboratory system for element specific hyperspectral X-ray imaging. *The Analyst*, **138 (3)**, 755–759.

Jonas, P. & Louis, A.K. (2004) Phase contrast tomography using holographic measurements. *Inverse Problems*, **20 (1)**, 75–102.

Kalender, W. (2006) X-ray computed tomography. *Physics in Medicine and Biology*, **51 (13)**, R29–R43.

Kallmann, H. (1948) Neutron radiography. *Research*, **1 (6)**, 254–260.

Kamenz, C., Dunlop, J.A., Scholtz, G., et al. (2008) Microanatomy of early Devonian book lungs. *Biology Letters*, **4 (2)**, 212–215.

Kardjilov, N., Manke, I., Hilger, A., et al. (2011) Neutron imaging in materials science. *Materials Today*, **14 (6)**, 248–256.

Keevil, S.F. (2001) Magnetic resonance imaging in medicine. *Physics Education*, **36 (6)**, 476–485.

Ketcham, R.A. & Carlson, W.D. (2001) Acquisition, optimization and interpretation of X-ray computed tomographic imagery: applications to the geosciences. *Computers & Geosciences*, **27 (4)**, 381–400.

Keyriläinen, J., Bravin, A., Fernández, M., et al. (2010) Phase-contrast X-ray imaging of breast. *Acta Radiologica*, **51 (8)**, 866–884.

Kinney, J.H. & Nichols, M.C. (1992) X-ray tomographic microscopy (XTM) using synchrotron radiation. *Annual Review of Materials Science*, **22 (1)**, 121–152.

Krug, K., Dik, J. & Leeuw, M. (2006) Visualization of pigment distributions in paintings using synchrotron K-edge imaging. *Applied Physics A*, **251**, 247–251.

Kruta, I., Landman, N., Rouget, I., et al. (2011) The role of ammonites in the Mesozoic marine food web revealed by jaw preservation. *Science*, **331**, 70–72.

Kuebler, K., McSween Jr., H.Y., Carlson, W.D., et al. (1999) Sizes and masses of chondrules and metal–troilite grains in ordinary chondrites: possible implications for nebular sorting. *Icarus*, **106**, 96–106.

Laaß, M., Hampe, O., Schudack, M., et al. (2011) New insights into the respiration and metabolic physiology of *Lystrosaurus*. *Acta Zoologica*, **92 (4)**, 363–371.

Lauterbur, P.C. (1973) Image formation by induced local interactions: examples employing nuclear magnetic resonance. *Nature*, **242 (5394)**, 190–191.

Legg, D.A., Garwood, R.J., Dunlop, J.A., et al. (2012) A taxonomic revision of Orthosternous scorpions from the English Coal-Measures aided by X-ray microtomography. *Palaeontologia Electronica*, **15 (2)**, 15.2.14A.

Lombardo, J.J., Ristau, R.A., Harris, W.M., et al. (2012) Focused ion beam preparation of samples for X-ray nanotomography. *Journal of Synchrotron Radiation*, **19** (5), 789–796.

Maisey, J.G. (2001) CT-scan reveals new cranial features in Devonian chondrichthyan *"Cladodus" wildungensis*. *Journal of Vertebrate Paleontology*, **21** (4), 807–810.

Marino, L., Uhen, M.D., Pyenson, N.D., et al. (2003) Reconstructing cetacean brain evolution using computed tomography. *Anatomical Record. Part B, New Anatomist*, **272** (1), 107–117.

Martí Mus, M. & Moczydłowska, M. (2000) Internal morphology and taphonomic history of the Neoproterozoic vase-shaped microfossils from the Visings Group, Sweden. *Norsk Geologisk Tidsskrift*, **80** (3), 213–228.

Mayo, S.C., Stevenson, A. & Wilkins, S. (2012) In-line phase-contrast X-ray imaging and tomography for materials science. *Materials*, **5**, 937–965.

Mamin, H.J., Kim, M., Sherwood, M.H., et al. (2013) Nanoscale nuclear magnetic resonance with a nitrogen-vacancy spin sensor. *Science*, **339** (6119), 557–560.

Mansfield, P., Pykett, I.L., Morris, P.G., et al. (1978) Human whole body line-scan imaging by NMR. *British Journal of Radiology*, **51** (611), 921–922.

McJury, M., Clark, N.D.L., Liston, J., et al. (1994) Dinosaur egg structure investigated using MRI. *Proceedings of the International Society for Magnetic Resonance in Medicine*, **1994** (S2), 706.

McMillan, E.M. (1945) The synchrotron – a proposed high energy accelerator. *Physics Review*, **68**, 143.

Mietchen, D., Keupp, H., Manz, B., et al. (2005) Non-invasive diagnostics in fossils – magnetic resonance imaging of pathological belemnites. *Biogeosciences*, **2** (2), 133–140.

Mietchen, D., Aberhan, M., Manz, B., et al. (2008) Three-dimensional magnetic resonance imaging of fossils across taxa. *Biogeosciences*, **5** (1), 25–41.

Minsky, M. (1961) US Patent 3013467.

Minsky, M. (1988) Memoir on inventing the confocal scanning microscope. *Scanning*, **10** (4), 128–139.

Molineux, A., Scott, R.W., Ketcham, R.A., et al. (2007) Rudist taxonomy using X-ray computed tomography. *Palaeontologia Electronica*, **10** (3), 13A, 6p.

Muehleman, C., Fogarty, D., Reinhart, B., et al. (2010) In-laboratory diffraction-enhanced X-ray imaging for articular cartilage. *Clinical Anatomy* **23** (5), 530–538.

Natterer, F. & Ritman, E.L. (2002) Past and future directions in X-ray computed tomography (CT). *International Journal of Imaging Systems and Technology*, **12** (4), 175–187.

Nix, T. & Feist-Burkhardt, S. (2003) New methods applied to the microstructure analysis of Messel oil shale: confocal laser scanning microscopy (CLSM) and environmental scanning electron microscopy (ESEM). *Geological Magazine*, **140** (4), 469–478.

O'Connor, B. (1996) Confocal laser scanning microscopy: a new technique for investigating and illustrating fossil radiolarian. *Micropaleontology*, **42** (4), 395–402.

Paganin, D., Mayo, S.C., Gureyev, T.E., et al. (2002) Simultaneous phase and amplitude extraction from a single defocused image of a homogeneous object. *Journal of Microscopy*, **206** (1), 33–40.

Penney, D., McNeil, A., Green, D., et al. (2011) A new species of anapid spider (Araneae: Araneoidea, Anapidae) in Eocene Baltic amber, imaged using phase contrast X-ray computed micro-tomography. *Zootaxa*, **66**, 60–66.

Peter, O. (1946) Neutronen-Durchleuchtung. *Zeitschrift für Naturforschung*, **1 (10)**, 551–559.

Petráň, M., Hadravský, M., Egger, M.D., et al. (1968) Tandem-scanning reflect-light microscope. *Journal of the Optical Society of America*, **58 (5)**, 661–664.

Petrik, V., Apok, V., Britton, J.A., et al. (2006) Godfrey Hounsfield and the dawn of computed tomography. *Neurosurgery*, **58 (4)**, 780–787.

Petrovic, A.M., Siebert, J.E. & Rieke, P.E. (1982) Soil bulk density analysis in three dimensions by computed tomographic scanning. *Soil Science*, **46**, 445–450.

Proussevitch, A., Ketcham, R.A., Carlson, W.D., et al. (1998) Preliminary results of X-ray CT analysis of Hawaiian vesicular basalts. *Eos*, **79 (17)**, s360.

Purcell, E.M., Torrey, H.C. & Pound, R.V. (1946) Resonance absorption by nuclear magnetic movements in a solid. *Physical Review*, **69 (1–2)**, 37–38.

Rabi, I.I., Zacharias, J.R., Millman, S., et al. (1938) A new method of measuring nuclear magnetic moment. *Physical Review*, **53 (4)**, 318.

Radon, J. (1917) Über die Bestimmung von Funktionen durch ihre Integralwerte längs gewisser Mannigfaltigkeiten. *Berichte über die Verhandlungen der Sächsische Akademie der Wissenschaften*, **69**, 262–277.

Rahman, I.A. & Zamora, S. (2009) The oldest cinctan carpoid (stem-group Echinodermata), and the evolution of the water vascular system. *Zoological Journal of the Linnean Society*, **157 (2)**, 420–432.

Rashid-Farrokhi, F., Liu, K.J.R., Berenstein, C.A., et al. (1997) Wavelet-based multiresolution local tomography. *IEEE Transactions on Image Processing*, **6 (10)**, 1412–1430.

Rau, C. & Somogyi, A. (2002) XANES micro-imaging and tomography. In: Bonse, U. (ed), *Developments in X-Ray Tomography III*, pp. 249–255. SPIE, Bellingham.

Rau, C., Somogyi, A. & Simionovici, A. (2003) Microimaging and tomography with chemical speciation. *Methods in Physics Research Section B*, **200**, 444–450.

Renter, J. (1989) Applications of computerized tomography in sedimentology. *Marine Geotechnology*, **8 (3)**, 201–211.

Ritman, E.L. (2004) Micro-computed tomography – current status and developments. *Annual Review of Biomedical Engineering*, **6**, 185–208.

Rogers, S.W. (1998) Exploring dinosaur neuropaleobiology: Viewpoint computed tomography scanning and analysis of an *Allosaurus fragilis* endocast. *Neuron*, **21 (4)**, 673–679.

Rowe, T.B. (1996) Coevolution of the mammalian middle ear and neocortex. *Science*, **273**, 651–654.

Rowe, T.B., Ketcham, R.A., Denison, C., et al. (2001) Forensic palaeontology: the Archaeoraptor forgery. *Nature*, **410**, 539–540.

Rüegsegger, P., Koller, B. & Müller, R. (1996) A microtomographic system for the nondestructive evaluation of bone architecture. *Calcified Tissue International*, **58 (1)**, 24–29.

Sanders, R. & Smith, D. (2005) The endocranium of the theropod dinosaur *Ceratosaurus* studied with computed tomography. *Acta Palaeontologica Polonica*, **50 (3)**, 601–616.

Sanderson, M.J. & Parker, I. (2003) Video-rate confocal microscopy. *Methods in Enzymology*, **360**, 447–481.

Schaller, S., Flohr, T., Klingenbeck, K., et al. (2000) Spiral interpolation algorithm for multislice spiral CT – part I: theory. *IEEE Transactions on Medical Imaging*, **19 (9)**, 822–834.

Schambach, S.J., Bag, S., Schilling, L., et al. (2010) Application of micro-CT in small animal imaging. *Methods*, **50 (1)**, 2–13.

Schillinger, B., Blümlhuber, W., Fent, A., et al. (1999) 3D neutron tomography: recent developments and first steps towards reverse engineering. *Nuclear Instruments and Methods in Physics Research A*, **424 (1)**, 58–65.

Schopf, J.W. & Kudryavtsev, A.B. (2009) Confocal laser scanning microscopy and Raman imagery of ancient microscopic fossils. *Precambrian Research*, **171 (1–2)**, 39–49.

Schopf, J.W. & Kudryavtsev, A.B. (2011) Confocal laser scanning microscopy and Raman (and fluorescence) spectroscopic imagery of permineralized Cambrian and Neoproterozoic fossils. In: Laflamme, M., Schiffbauer, J.D. & Dornbos, S.Q. (eds), *Quantifying the Evolution of Early Life: Numerical Approaches to the Evaluation of Fossils and Ancient Ecosystems*, pp. 241–270. Springer, Dordrecht.

Schopf, J.W., Tripathi, A.B. & Kudryavtsev, A.B. (2006) Three-dimensional confocal optical imagery of Precambrian microscopic organisms. *Astrobiology*, **6 (1)**, 1–16.

Schopf, J.W., Tewari, V.C. & Kudryavtsev, A.B. (2008) Discovery of a new chert-permineralized microbiota in the Proterozoic Buxa formation of the Ranjit Window, Sikkim, northeast India, and its astrobiological implications. *Astrobiology*, **8 (4)**, 735–746.

Schwarz, D., Vontobel, P., Lehmann, E.H., et al. (2005) Neutron tomography of internal structures of vertebrate remains: a comparison with X-ray computed tomography. *Palaeontologia Electronica*, **8 (2)**, 30A.

Scheyer, T. (2008) Aging the oldest turtles: the placodont affinities of *Priscochelys hegnabrunnensis*. *Naturwissenschaften*, **95 (9)**, 803–810.

Scott, A.C. & Hemsley, A.R. (1991) A comparison of new microscopical techniques for the study of fossil spore wall ultrastructure. *Review of Palaeobotany and Palynology*, **67 (1–2)**, 133–139.

Seet, K.Y.T., Barghi, A., Yartsev, S., et al. (2009) The effects of field-of-view and patient size on CT numbers from cone-beam computed tomography. *Physics in Medicine and Biology*, **54 (20)**, 6251–6262.

Selden, P.A., Shear, W.A. & Sutton, M.D. (2008) Fossil evidence for the origin of spider spinnerets, and a proposed arachnid order. *Proceedings of the National Academy of Sciences of the United States of America*, **105 (52)**, 20781–20785.

Siegmund, O.H.W., Vallerga, J.V., Martin, A., et al. (2007) A high spatial resolution event counting neutron detector using microchannel plates and cross delay line readout. *Nuclear Instruments and Methods in Physics Research A*, **579 (1)**, 188–191.

Smith, T.B. & Nayak, K.S. (2010) MRI artifacts and correction strategies. *Imaging in Medicine*, **2 (4)**, 445–457.

Snigirev, A. & Snigireva, I. (1995) On the possibilities of X-ray phase contrast microimaging by coherent high-energy synchrotron radiation. *Review of Scientific Instruments*, **66 (12)**, 5486– 5492.

Speranza, M., Wierzchos, J., Alonso, J., et al. (2010) Traditional and new microscopy techniques applied to the study of microscopic fungi included in amber. In: Méndez-Vilas, A. & Díaz, J. (eds), *Microscopy: Science, Technology, Applications and Education*, pp. 1135–1145. Formatex, Badajoz.

Stampanoni, M, Borchert, G., Abela, R., et al. (2003) Nanotomography based on double asymmetrical Bragg diffraction. *Applied Physics Letters*, **82 (17)**, 2922–2294.

Steiger, T. (2001) Nuclear magnetic resonance imaging in paleontology. *Computers & Geosciences*, **27 (4)**, 493–495.

Strobl, M., Manke, I., Kardjilov, N., et al. (2009) Advances in neutron radiography and tomography. *Journal of Physics D: Applied Physics*, **42 (24)**, 243001.

Svishchev, G.M. (1969) Microscope for the study of transparent light-scattering objects in incident light. *Optics and Spectroscopy*, **26 (2)**, 171–172.

Tafforeau, P., Boistel, R., Boller, E., et al. (2006) Applications of X-ray synchrotron microtomography for non-destructive 3D studies of paleontological specimens. *Applied Physics A*, **83 (2)**, 195–202.

Talyzina, N.M. (1998) Fluorescence intensity in Early Cambrian acritarchs from Estonia. *Review of Palaeobotany and Palynology*, **100 (1–2)**, 99–108

Tate, J.R. & Cann, C.E. (1982) High-resolution computed tomography for the comparative study of fossil and extant bone. *American Journal of Physical Anthropology*, **58 (1)**, 67–73.

Thewlis, J. (1956) Neutron radiography. *British Journal of Applied Physics*, **7 (10)**, 345–350.

Thompson, J.L. & Illerhaus, B. (1998) A new reconstruction of the Le Moustier 1 skull and investigation of internal structures using 3-D-μCT data. *Journal of Human Evolution*, **35 (6)**, 647–665.

Tremsin, A.S., Vallerga, J.V., McPhate, J.B., et al. (2008) On the possibility to image thermal and cold neutron with sub-15 μm spatial resolution. *Nuclear Instruments and Methods in Physics Research A*, **592 (3)**, 374–384.

Tyszka, J.M., Fraser, S.E. & Jacobs, R.E. (2005) Magnetic resonance microscopy: recent advances and applications. *Current Opinion in Biotechnology*, **16 (1)**, 93–99.

Veksler, V. (1944) A new method of accelerating relativistic particles. *Comptes rendus de l'Académie des Sciences de l'URSS*, **8**, 329–331.

Vendrasco, M.J., Wood, T.E. & Runnegar, B.N. (2004) Articulated Palaeozoic fossil with 17 plates greatly expands disparity of early chitons. *Nature*, **429**, 288–291.

Vlassenbroeck, J., Cnudde, V., Masschaele, B., et al. (2007) A comparative and critical study of X-ray CT and neutron CT as non-destructive material evaluation techniques. *Geological Society, London, Special Publications*, **271**, 277–285.

Vontobel, P., Lehmann, E.H., Hassanein, R., et al. (2006) Neutron tomography: method and applications. *Physica B*, **385–386 (1)**, 475–480.

Wang, G., Vannier, M.W. & Cheng, P.C. (1999) Iterative X-ray cone-beam tomography for metal artifact reduction and local region reconstruction. *Microscopy and Microanalysis*, **5 (1)**, 58–65.

Webb, S. (1990) *From the Watching of Shadows: The Origins of Radiological Tomography*. Taylor & Francis, London.

Weitkamp, T. & Haas, D. (2011) ANKAphase: software for single-distance phase retrieval from inline X-ray phase-contrast radiographs. *Journal of Synchrotron Radiation*, **18**, 617–629.

Wellington, S.L. & Vinegar, H.J. (1987) X-ray computerized tomography. *Journal of Petroleum Technology*, **39 (8)**, 885–898.

White, J.G. & Amos, W.B. (1987) Confocal microscopy comes of age. *Nature*, **328 (6126)**, 183–184.

Wilkins, S.W., Gureyev, T. & Gao, D. (1996) Phase-contrast imaging using polychromatic hard X-rays. *Nature*, **384**, 335–338.

Wind, J. (1984) Computerized X-ray tomography of fossil hominid skulls. *American Journal of Physical Anthropology*, **63 (3)**, 265–282.

Wind, J. & Zonneveld, F.W. (1989) Computed tomography of an *Australopithecus* skull (Mrs Ples): a new technique. *Naturwissenschaften*, **327**, 325–327.

Winkler, B. (2006) Applications of neutron radiography and neutron tomography. *Reviews in Mineralogy and Geochemistry*, **63 (1)**, 459–471.

Withers, P.J. (2007) X-ray nanotomography. *Materials Today*, **10 (12)**, 26–34.

Witzmann, F., Scholz, H., Müller, J., et al. (2010) Sculpture and vascularization of dermal bones, and the implications for the physiology of basal tetrapods. *Zoological Journal of the Linnean Society*, **160 (2)**, 302–340.

Yang, J., Regier, T., Dynes, J.J., et al. (2011) Soft X-ray induced photoreduction of organic Cu(II) compounds probed by X-ray absorption near-edge (XANES) spectroscopy. *Analytical Chemistry*, **83 (20)**, 7856–7862.

Ziegler, A., Kunth, M., Mueller, S., et al. (2011) Application of magnetic resonance imaging in zoology. *Zoomorphology*, **130 (4)**, 227–254.

Zonneveld, F.W., Spoor, C.F. & Wind, J. (1989) The use of computed tomography in the study the internal morphology of hominid fossils. *Medicamundi*, **34**, 117–128.

Zonneveld, F.W. & Wind, J. (1985) High-resolution computed tomography of fossil hominid skulls: a new method and some results. In: Tobias, P.V. (ed), *Hominid Evolution: Past, Present and Future*, pp. 427–436. Wiley-Blackwell, Malden.

Further Reading/Resources

Abel, R.L., Laurini, C. & Richter, M. (2012) A palaeobiologist's guide to "virtual" micro-CT preparation. *Palaeontologia Electronica*, **15** (2), 6T, 17p.

Als-Nielsen, J. & McMorrow, D. (2011) *Elements of Modern X-Ray Physics*. Wiley-Blackwell, Malden.

Anderson, I.S., McGreevy, R. & Bilheux, H.Z. (eds) (2009) *Neutron Imaging and Applications: A Reference for the Imaging Community*. Springer, New York.

Callaghan, P.T. (1991) *Principles of Nuclear Magnetic Resonance Microscopy*. Clarendon Press, Oxford.

Carlson, W.D. (2006) Three-dimensional imaging of earth and planetary materials. *Earth and Planetary Science Letters*, **249 (3–4)**, 133–147.

Dieing, T., Hollricher, O. & Toporski, J. (eds) (2011) *Confocal Raman Microscopy*. Springer, Heidelberg.

Domanus, J.C. (1992) *Practical Neutron Radiography*. Kluwer Academic Publishers, Dordrecht.

Haacke, E.M., Brown, R.W., Thompson, M.R., et al. (1999) *Magnetic Resonance Imaging: Physical Principles and Sequence Design*. Wiley-Blackwell, New York.

Haisch, C. (2012) Optical tomography. *Annual Review of Analytical Chemistry*, **5**, 57–77.

Hsieh, J. (2003) *Computed Tomography: Principles, Design, Artifacts, and Recent Advances*. Wiley Inter-Science/SPIE Press, Bellingham.

Jacques, S.D.M., Egan, C.K., Wilson, M.D., et al. (2013) A laboratory system for element specific hyperspectral X-ray imaging. *The Analyst*, **138 (3)**, 755–759.

Kalender, W. (2006) X-ray computed tomography. *Physics in Medicine and Biology*, **51 (13)**, R29–R43.

Liang, L., Rinaldi, R. & Schober, H. (eds) (2009) *Neutron Applications in Earth, Energy and Environmental Sciences*. Springer, New York.

Morris, P.G. (1986) *Nuclear Magnetic Resonance Imaging in Medicine and Biology*. Clarendon Press, Oxford.

Pawley, J.B. (ed) (2006) *Handbook of Biological Confocal Microscopy*. Springer, Berlin.

Sheppard, C.J.R. & Shotton, D.M. (1997) *Confocal Laser Scanning Microscopy*. Springer, New York.

Vogel, S.C. & Priesmeyer, H.-G. (2006) Neutron production, neutron facilities and neutron instrumentation. *Reviews in Mineralogy and Geochemistry*, **63 (1)**, 27–57.

Withers, P.J. (2007) X-ray nanotomography. *Materials Today*, **10 (12)**, 26–34.

4

Surface-Based Methods

Abstract: Surface-based methods are used to digitize the three-dimensional surfaces of fossils. Laser scanning is the most widespread technique; here, laser light is used to actively probe a target at distance. Photogrammetry typically involves the passive detection of visible light, for example through photography of an object from multiple viewpoints. Neither approach requires contact with the sample, and both are capable of producing high-resolution colour datasets; photogrammetry is substantially cheaper and quicker than laser scanning. A third method, mechanical digitization, uses a mechanical arm to probe a surface through physical touch. This technique is less accurate and more labour intensive than laser scanning and photogrammetry. To date, all three methods have been predominantly used to study fossil vertebrates; however, laser scanning and photogrammetry, in particular, are likely to be applicable to a wide range of preservation types, taxonomic groups and specimen sizes.

4.1 Introduction

Surface-based methods non-destructively capture the topography (i.e. surface features) of an object in three dimensions. The collected datasets are not volumetric, that is, they do not include details of the sample's interior. Volumetric information can be critical to researchers, especially when a fossil is embedded in rock matrix, or preserves complex internal anatomy (see Chapters 2 and 3 for numerous examples), but is not necessarily a prerequisite of work with virtual fossils. Some studies will require access only to the three-dimensional form of the external surface of a specimen, especially when the goal is to use this surface in additional quantitative or modelling analyses (see Chapter 6). Some surface-based techniques can also capture data on the physical appearance (e.g. colour) of samples. These details can be extremely valuable for visualization, dissemination and display purposes.

Techniques for Virtual Palaeontology, First Edition. Mark D. Sutton, Imran A. Rahman and Russell J. Garwood.
© 2014 John Wiley & Sons, Ltd. Published 2014 by John Wiley & Sons, Ltd.

The methods described in this chapter are relatively inexpensive, accessible, non-destructive and rapid – this combination of traits cannot be found in any single tomographic technique (although X-ray microtomography comes close). Surface-based methods, therefore, represent powerful tools in the virtual palaeontologist's armoury. We detail three techniques, differing in the approach they take to imaging surfaces: laser scanning (Section 4.2), photogrammetry (Section 4.3) and mechanical digitization (Section 4.4).

4.2 Laser Scanning

Laser scanning is the most common surface-based technique employed today. The method uses a laser beam to characterize the exterior three-dimensional shape and appearance of an object (or area) at distance. A number of different scanners are available, from handheld devices optimized for imaging smaller samples at sub-millimetre resolutions, to long-range systems (e.g. light detection and range, or LiDAR) capable of scanning larger specimens or field sites at resolutions down to a few millimetres. Laser scanning has found widespread use in palaeontology since the 2000s (e.g. Lyons et al. 2000; Béthoux et al. 2004; Bates et al. 2008; Antcliffe and Brasier 2011).

4.2.1 History

The history of laser science dates back to the mid-1910s, when Einstein theorized the concept of stimulated emission (Einstein 1916, 1917), the process by which electromagnetic waves (of the appropriate frequency) can stimulate an excited atom or molecule to transition to a lower energy state and emit further waves of the same frequency. The first functional laser – light amplification by stimulated emission of radiation – was reported several decades later (Maiman 1960). This instrument used pulses of light to excite a ruby crystal and thereby produce flashes of red laser light. Following further technical advancements in the 1960s, lasers gradually began to find practical applications, notably for scanning through the controlled deflection of laser beams. Large-scale laser range-finders were first produced in the 1960s and 1970s, and were initially used for atmospheric monitoring and terrain mapping in three dimensions (Schuster 1970; Jarvis 1983). Slightly later, the first three-dimensional object scanners were produced (e.g. Addleman and Addleman 1985; Arridge et al. 1985). Early commercial applications included digitizing human faces (with eye-safe lasers) for use in films such as *The Abyss* and *Terminator 2*. Laser-scanning techniques have developed considerably since, their size and cost decreasing, and their speed and accuracy increasing. Laser-scanning technology is now mature and well commercialized, and the instruments are used for research in a diverse range of fields including medicine, civil engineering, archaeology and palaeontology.

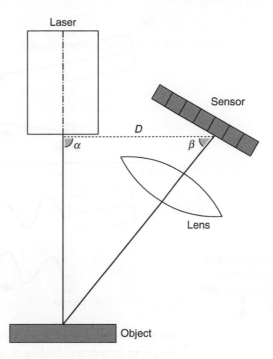

Figure 4.1 Triangulation-based laser scanning. *Abbreviations*: D, distance between the laser source and the sensor; α, angle of the source to the object and β, angle of the sensor to the object.

4.2.2   *Principles and Practicalities*

Laser scanners use laser light to acquire three-dimensional positions of points on a surface without direct physical contact; together, these points (defined as x-, y- and z-coordinates) define a **point cloud**. If scanning is combined with digital photography, information about the colour and surface texture of the object can be associated with individual points within the cloud. Three main classes of laser scanner are discussed in the following text: triangulation, time-of-flight and phase shift.

4.2.2.1   Triangulation

Triangulation-based scanners work by projecting a spot or strip of laser light onto an object, with the reflected light collected by a lens and received by a sensor (typically a charge-coupled device or positron-sensitive detector) offset from the source. The distance between the laser source and the sensor (D), and the angles of the source (α) and the sensor (β) to the object (Figure 4.1) are used to determine the position of the illuminated point(s) on the surface through triangulation. In some systems, the laser beam is automatically scanned across the object's surface during data capture. Alternatively, full coverage can be achieved by moving the object (e.g. using a rotation stage) or the scanner (e.g. with a mechanical arm or using a hand-held device); this enables digitization of the whole specimen (including rear, upper and lower surfaces). A variety of instruments exist, suitable for imaging surfaces at resolutions down to about 50 μm, and/or at ranges of up to a

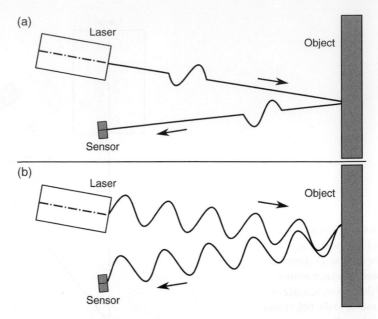

Figure 4.2 Comparison of (a) time-of-flight laser scanning and (b) phase-shift laser scanning.

few metres. Triangulation scanners often perform poorly in bright sunlight, requiring temporary shading of the object to be imaged (Remondino et al. 2010). Portable, handheld devices are particularly useful in palaeontology, with fields of view that are appropriate for bones of medium-sized vertebrates and whole invertebrates (i.e. surfaces < 1 m in size). Larger objects can potentially be scanned in multiple parts, but this can be very time consuming (e.g. weeks for a complete vertebrate skeleton) and necessitates data capture at lower, millimetre-scale resolutions (otherwise datasets will be too large for even high-end computers). **Triangulation-based laser scanning** has been used to study a range of different fossils, including the Ediacara biota (Brasier and Antcliffe 2009; Antcliffe and Brasier 2011), insects (e.g. Béthoux et al. 2004), trace fossils (e.g. Petti et al. 2008; Platt et al. 2010) and vertebrates (Lyons et al. 2000; Zhang et al. 2000).

4.2.2.2 Time-of-Flight

Time-of-flight scanners emit a short pulse of laser light (Figure 4.2a), which is reflected off the surface of the object and detected with a sensor (e.g. an avalanche photodiode). Light takes approximately 3.33 picoseconds to travel 1 mm; the time taken for the light to return to the scanner is measured, divided by two and multiplied by the speed of light to determine the distance between the scanner and a point on the surface of interest. This is repeated for the entire surface by rapidly scanning it with the laser beam (typically using a system of rotating mirrors), capturing tens of thousands of coordinate points every second. The intensity of the laser return can be recorded at the same time, providing additional textural information (e.g. target reflectivity). Terrestrial time-of-flight scanners are capable of imaging objects at resolutions down to a few millimetres and can operate at long ranges, perhaps a kilometre or more (airborne time-of-flight scanners work at even longer ranges). These ground-based units are semi-portable (i.e. mounted on a tripod, battery powered and operated from a laptop), have very large fields of view (e.g. 180° vertical and 360° horizontal) and perform well under demanding environmental conditions. They are, therefore, well suited for digitizing large surfaces in non-laboratory settings (Bellian et al. 2005). In palaeontology, time-of-flight laser scanners have been used to study large tracksites (e.g. Breithaupt et al. 2004; Bates et al. 2008) and whole large vertebrate skeletons (e.g. Bates et al. 2009a).

4.2.2.3 Phase Shift

Phase-shift scanners operate in a similar way to time-of-flight scanners, but instead emit a laser beam with sinusoidally modulated optical power (Figure 4.2b), and measure the distance from the scanner to the object by comparing the change in phase between the emitted and reflected laser light. The rate of data capture is substantially higher than other laser-scanning methods, with hundreds of thousands of coordinate points acquired per second, producing high-density point clouds very quickly. Like time-of-flight scanners, phase-shift scanners have very large fields of view, offer resolutions down to a few millimetres and are semi-portable; however, they operate at shorter ranges, typically up to about 100 m. **Phase-shift laser scanning** is suitable for rapidly imaging large surfaces at relatively short ranges in the field; it has been used only rarely in palaeontology, for example to study mass accumulations of ammonites (Lukeneder and Lukeneder 2011) and bivalves (Haring et al. 2009).

4.2.2.4 General Considerations

Laser scanning does not normally require any sample preparation (although highly reflective surfaces may be easier to image if coated with e.g. baby powder – see Zhang et al. 2000), and the lasers used are generally eye safe, so additional safety precautions are not required. The optimal scanning approach – triangulation, time-of-flight, or phase shift – will depend on the scale of the object and the desired resolution, the distance between the scanner and the object, and any restrictions on scanning time (as well as instrument availability). Samples less than about 1 m in size are best imaged using triangulation-based scanners; such units are capable of the highest resolutions of the laser-scanning methods discussed herein. Larger objects could potentially also be captured with triangulation laser scanners by acquiring multiple scans; however, this would result in very long data-capture times and lower spatial resolutions (owing to constraints on usable dataset sizes). Time-of-flight and phase-shift laser scanning are more appropriate techniques for the study of large (>1 m) surfaces such as fossil tracksites. Both these methods are capable of millimetre-scale resolutions and are semi-portable (enabling their use in non-laboratory settings). The main differences between the two approaches concern their range and speed: time-of-flight scanners are capable of imaging at much greater distances, but require longer data-acquisition times; phase-shift scanners have shorter ranges, but scans are more rapid.

Selecting the optimal surface coverage (i.e. how many point coordinates are needed to accurately represent the smallest feature of interest) is another important consideration in laser scanning. This is essentially a trade-off between data capture time and accuracy – longer scans allow for the capture of more data points – assuming the scanner used is capable of sufficiently accurate and precise measurements. In practice, it can be difficult to determine the appropriate point density prior to a scan; monitoring the growing point cloud on a computer during scanning is the best way to establish this value.

If multiple scans are carried out for a single specimen – for example, to capture all surfaces of the object, or to ensure full coverage of a very large surface – the resulting point clouds will need to be registered using computer software to obtain a complete three-dimensional reconstruction (see Section 5.3). This can often be achieved without any external calibration, but reference objects (such as spheres) are sometimes incorporated into the scanned scenes to aid later **manual registration** (see e.g. Stoinski 2011).

4.2.3   Case Studies of Methodology

4.2.3.1   Triangulation-Based Laser Scanning

In a series of publications, Antcliffe and Brasier used laser scanning to characterize the morphology of several Ediacaran fossils (e.g. Antcliffe and Brasier 2008, 2011; Brasier and Antcliffe 2009). These fossils are difficult to photograph owing to their low relief, vary in size from less than 1 cm to several centimetres in length, and in many cases must be studied *in situ* (often at remote field localities). Therefore, a portable triangulation-based laser scanner was selected for this work. Using this instrument, fossils were imaged at resolutions down to 60 µm. The resulting point clouds were visualized using the software Demon3D (www.archaeoptics.co.uk/products/demon3d/); three-dimensional surfaces were illuminated at different angles with virtual coloured light, thereby revealing subtle morphological features and avoiding issues with ambient lighting traditional in the study of these fossils (Figure 4.3).

4.2.3.2   Time-of-Flight Laser Scanning

Laser scanning has proven particularly valuable for the three-dimensional documentation of large bedding planes preserving trace fossils. For example, Bates et al. (2008) used the technique to survey the Fumanya dinosaur tracksites in north-east Spain. Here, footprints occur across several hundreds of metres of rock; this scale necessitated the use of **time-of-flight laser scanning**. This study utilized a RIEGL LMS-Z420i LiDAR scanner with a range of 800 m and a field of view of 80° (vertical) × 360° (horizontal). The scanner was coupled with a digital camera and a laptop, and mounted on a tripod. 17 scan stations were used in order to eliminate shadows (i.e. areas not illuminated by the laser) and ensure full coverage of the outcrop (by integrating scans). First, a 360° panoramic scan was performed at each scan station, providing dense coverage

Figure 4.3 Three-dimensional surface of the Ediacaran fossil *Charnia masoni* produced using triangulation-based laser scanning. Arrows indicate four different directions of virtual lighting. Scale bar is 30 mm. Source: Brasier and Antcliffe (2009, Fig. 1).

Figure 4.4 Three-dimensional surfaces of the Fumanya South dinosaur tracksite produced using time-of-flight laser scanning. Surfaces textured with digital photographs. (a) Outcrop. Scale bar is 50 m. (b) Track-bearing surface. Scale bar is 2 m. Source: Bates et al. (2008, Fig. 6).

(1,998,000 points per scan) for the entire exposure. The track-bearing surface was then scanned at high resolution (point spacings of 10–80 mm) at each station. The point clouds resulting from scans were textured using digital photographs taken at the stations during optimal natural lighting conditions (Figure 4.4). Multiple scans were merged into a unified three-dimensional surface with the program PolyWorks (www.innovmetric.com). This enabled interactive three-dimensional visualization of the dinosaur tracksites.

4.3 Photogrammetry

Photogrammetry is a method for determining the three-dimensional surface topography of an object or area from multiple two-dimensional images acquired at different viewpoints. Typically, a digital camera is used to capture photographs of the target; this approach is scale less, and so is theoretically applicable to surfaces of all sizes. In palaeontology, photogrammetric techniques have hitherto primarily been used to document dinosaur skeletons (e.g. Wiedemann et al. 1999; Stoinski 2011) and tracksites (e.g. Breithaupt and Matthews 2001; Matthews et al. 2006; Bates et al. 2009b), but recent work has demonstrated the potential of the method for studying a much wider range of fossil taxa (Falkingham 2012).

4.3.1 History

Photogrammetry has an extensive history, going back almost as far as the invention of photography in the early to mid-1800s, and encompassing the development of stereoscopic viewing. Initially, it was predominantly used for mapping and aerial reconnaissance, notably during the First and Second World Wars. Subsequently, the advent of the computer dramatically improved the accuracy of the technique, bringing about more widespread usage. This culminated in the emergence of digital photogrammetry in the 1980s, following the invention of the digital camera. Modern applications include cartography, medicine, forensics, archaeology and palaeontology.

4.3.2 Principles and Practicalities

Photogrammetry involves the measurement of homologous points in two or more overlapping images (captured at different positions) in order to reconstruct three-dimensional coordinates of the surface of interest (i.e. a point cloud) through triangulation. The images used are typically acquired passively, that is, by capturing natural light with photography. They therefore incorporate information about the colour and texture of the imaged surface, which can be integrated directly into the point cloud and/or used in post-processing to photo-texture a three-dimensional mesh. Photogrammetry is often subdivided into aerial photogrammetry (which uses aerial cameras; object-to-camera distance > 300 m) and close-range photogrammetry (which uses handheld or tripod-mounted cameras; object-to-camera distance < 300 m). This section focuses on the second of these approaches, which is significantly cheaper, quicker, more accessible and resolves finer details than the first; it is suitable for most palaeontological specimens/sites.

Close-range photogrammetry normally requires only basic equipment: a digital camera, computer and appropriate software – differentiating it from aerial photogrammetry, for example, where an aircraft equipped with a long-range camera is necessary. The method is therefore highly portable, and can be undertaken in non-laboratory settings; however, consistent lighting conditions are desirable to enhance the quality of the final three-dimensional visualization. The choice of digital camera is not critical; modern low-cost consumer cameras – SLR, compact and smartphone – frequently have high pixel counts (e.g. >6 megapixels), and are capable of generating datasets with sub-millimetre resolution. The resolution of photogrammetry is closely linked to the number of images acquired; taking more photographs closer to the target will generally result in a higher-density point cloud, and hence better accuracy/coverage. Additionally, a high-quality macro lens could be beneficial for high-resolution imaging of smaller, centimetre-sized objects. Recent studies have shown that photogrammetry is capable of producing reconstructions of palaeontological material with similar or better resolutions (i.e. denser point clouds) than many laser scanners (Petti et al. 2008; Remondino et al. 2010; Falkingham 2012).

Digital photogrammetry is a very rapid data-capture methodology; acquiring photographs will typically be quicker than laser scanning a surface. The number of photographs required will depend not only on the resolution desired, but also the nature of the specimen being studied. For example, complex samples such as vertebrate skeletons require more images (up to hundreds) to ensure full coverage of all surfaces. However, the larger the number of images, the longer the acquisition time and the greater the computational resources needed to generate the point cloud. Ideally, photographs should overlap by about 66% (i.e. by taking photographs a set distance apart), with any given point present in at least three images (Falkingham 2012). If these conditions are met, photogrammetry can even be used to reconstruct archived photographs. In addition, the method lacks any inherent notion of scale, and so an object of known size (i.e. a scale bar) should be included in scenes, or part of the target measured, in order to correctly scale the digital reconstruction. A variety of commercial and open-source software packages are available for creating three-dimensional surfaces from photographic images. Reconstruction, which is now generally automated, involves locating matching points between photographs and aligning them to generate a point cloud consisting of hundreds of thousands, or even millions of points. The point cloud can be visualized directly or converted into a **triangle mesh** (see Section 5.3).

Although theoretically applicable at all scales, in practice, photogrammetry with a digital camera does not work well for specimens smaller than about 10 mm (Falkingham, personal communication, 2013). An alternative approach is to use a microscope; for example, scanning electron microscope (SEM) photogrammetry is suitable for imaging very small (e.g. <1 mm) samples at nanometre-scale resolutions (Piazzesi 1973). Here, the specimen is imaged at two different tilt angles (a few degrees apart), with the resulting micrographs used to create a three-dimensional digital elevation model. This approach has proven useful for the study of microscopic craters (e.g. Kearsley et al. 2007) and could also be appropriate for microfossils, although no palaeontological work that utilizes it has yet been published.

4.3.3 Case Study of Methodology

Falkingham (2012) used photogrammetry to digitize a variety of different specimens, ranging from ~40 mm to 50 m in size, and thereby establishing the broad utility of the technique in palaeontology. This study used an Olympus E-500 8-megapixel digital camera to acquire images of a trilobite (35 photographs), a *Chirotherium* trackway (50 photographs), a tree root system (24 photographs), a mounted elephant skeleton (two datasets; 44 photographs and 207 photographs) and the front of the Manchester Museum (Manchester, UK) (52 photographs). The number of photographs was adjusted according to the complexity of the samples (two datasets were acquired for the elephant skeleton to test the influence of photograph number on point-cloud quality), with photographs taken 15° apart or less. Point clouds were produced for all these objects in Linux using the free software Bundler (phototour.cs.washington.edu/bundler) and associated programs. The resulting point clouds varied in density from 179,294 points (trilobite) to 2,171,040 (*Chirotherium*), and in all cases recorded key surface features of the objects. Furthermore, a cast of a bird track was imaged with both photogrammetry (75 photographs) and triangulation-based laser scanning (resolution of ~0.3 mm) to compare these two surface-based methods. Here, the photogrammetric dataset generated a much higher-density point cloud, revealing subtle surface details not apparent in the laser-scan data (Figure 4.5). Note that since Falkingham's study (2012), the software packages have become streamlined and easier to use, in the form of VisualSFM (homes.cs.washington.edu/~ccwu/vsfm).

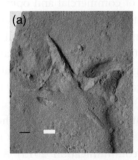 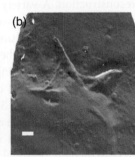 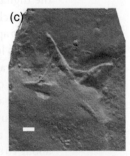

Figure 4.5 Cast of a bird track. (a) Photograph of the original specimen. (b) Three-dimensional surface produced using laser scanning. (c) Three-dimensional surface produced using photogrammetry. Scale bars are 10 mm. Source: Falkingham (2012, Fig. 10). Reproduced with permission of the Palaeontological Association.

4.4 Mechanical Digitization

Mechanical or contact digitization uses a jointed mechanical arm which contains rotational and/or positional sensors at each joint, and bears a digitizing tip. The tip is moved automatically or, more normally, manually over the surface of a specimen, and its position in three dimensions is recorded by the sensors. The technique has been used in vertebrate palaeontology and palaeoanthropology for the collection of landmark data for morphometric studies (e.g. Bonnan 2004; Goswami 2004; Green and Alemseged 2012), but can also function as a data-capture methodology for virtual palaeontology (Wilhite 2003; Mallison et al. 2009).

A detailed description of the methodologies involved is provided by Mallison et al. (2009); we present a summary here. Data can be collected in one of two modes. Position measurements can be distributed over the entire surface of the fossil to generate a point cloud (Figure 4.6c, d). Alternatively, they can be made in sub-parallel closed loops around a specimen (Figure 4.6a), and subsequently reconstructed using vector-surfacing approaches (Section 5.2.3); Mallison et al. (2009) used non-uniform rational B-spline (NURBS) surfacing, performed in the software package Rhinoceros (www.rhino3D.com); see Figure 4.6d. The closed-loop digitization method generates models of a lower file size, and avoids the need for the problematic point-cloud triangulation stage (Section 5.3). It does, however, require that the specimen has loops (or at least start/end points of loops) manually marked onto it to guide digitization; these can be drawn onto transparent tape or any other overlay on the specimen. Whichever method is used, specimens can rarely be digitized without recalibration (i.e. moving the specimen and/or digitizer). This is normally required to enable rear and/or upper and lower surfaces to be reached, but potentially also to enable full

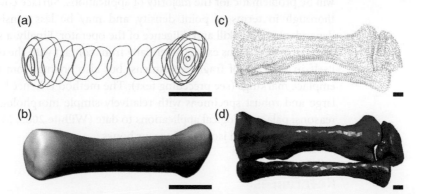

Figure 4.6 Dinosaur bones captured through mechanical digitization. (a) Closed-loop data for a *Plateosaurus engelhardti* left radius. (b) Vector-surfaced (NURBS) reconstruction of data from (a). (c) Point-cloud data for a *Dicraeosaurus sattleri* lower hind limb; the three bones are represented by point clouds of different colours. (d) Polygon-mesh reconstruction of data from (c). Scale bars are 50 mm. After Mallison et al. (2009, Figs. 2, 3). Reproduced with permission of the Society for Vertebrate Palaeontology.

coverage of a large specimen. To enable data from these separate sessions to be combined into a single dataset, calibration points must be marked on the specimen; thus, even where the point-cloud approach is to be used, emplacement of marks is difficult to avoid. Where loop-based digitization is employed, it is preferable to digitize entire closed loops; digitizing part loops (e.g. on each side of the specimen) and connecting them digitally at a later stage is possible, but time consuming and prone to error.

Prior to mechanical digitization, specimens must be immobilized, either using clamp arrangements, or by placing them in a sandbox or mounting in media such as clay. Minimum specimen size is primarily controlled by difficulties in immobilizing specimens below 50 mm (Mallison et al. 2009), and the loop-digitization variant is not appropriate for specimens below 100 mm. No theoretical upper size limit exists, and specimens greater than 2 m in size have been digitized (Mallison et al. 2009); however, the need for multiple recalibrations (see preceding text) can reduce accuracy in large specimens. Digitization accuracy under ideal conditions can reach 50 μm (manufacturer's data, www.3d-microscribe.com); Mallison et al (2009) analysed errors by comparing with computed tomography (CT) data, and found that accuracy was near to 1 mm (for a 200 mm specimen). They also noted that the point-cloud variant, in comparison to the loop-digitization variant, produces datasets with a higher average deviation, but a lower maximum deviation.

Mechanical digitization is relatively cheap, and data-capture time is normally low. The resultant datasets and reconstructions are also small, facilitating their dissemination, storage and visualization on low-powered computers. There are, however, several disadvantages to the method when compared to non-contact surface-based approaches (i.e. laser scanning and photogrammetry; see Sections 4.2 and 4.3). Its accuracy is generally lower and surface colour is not normally captured, although neither of these issues will be problematic for the majority of applications. Surface coverage is less thorough in terms of point density, and may be less consistent, being dependent on the skill and diligence of the operator. Finally, a small risk of damage to specimens exists, potentially from contact with the digitizing tip itself in the case of fragile specimens, but also arising from the need to emplace markings (see preceding text). The method is hence best suited to large and robust specimens with relatively simple morphology; for these reasons, palaeontological applications to date (Wilhite 2003; Mallison et al. 2009) have involved isolated dinosaur bones.

References

Addleman, D. & Addleman, L. (1985) Rapid 3D digitizing. *Computer Graphics World*, **8 (11)**, 41–44.

Antcliffe, J.B. & Brasier, M.D. (2008) *Charnia* at 50: developmental models for Ediacaran fronds. *Palaeontology*, **51 (1)**, 11–26.

Antcliffe, J.B. & Brasier, M.D. (2011) Fossils with little relief: using lasers to conserve, image, and analyse the Ediacara biota. In: Laflamme, M., Schiffbauer, J.D. & Dornbos, S.Q. (eds), *Quantifying the Evolution of Early Life: Numerical Approaches to the Evaluation of Fossils and Ancient Ecosystems*, pp. 223–240. Springer, Dordrecht.

Arridge, S., Moss, J.P., Linney, A.D. & James, D.R. (1985) Three dimensional digitization of the face and skull. *Journal of Maxillofacial Surgery*, **13 (3)**, 136–143.

Bates, K.T., Rarity, F., Manning, P.L., et al. (2008) High-resolution LiDAR and photogrammetric survey of the Fumanya dinosaur tracksites (Catalonia): implications for the conservation and interpretation of geological heritage sites. *Journal of the Geological Society, London*, **165 (1)**, 115–127.

Bates, K.T., Manning, P.L., Hodgetts, D. & Sellers, W.I. (2009a) Estimating mass properties of dinosaurs using laser imaging and 3D computer modelling. *PLoS ONE*, **4 (2)**, e4532.

Bates, K.T., Breithaupt, B.H., Falkingham, P.L., Matthews, N., Hodgetts, D. & Manning, P.L. (2009b) Integrated LiDAR and photogrammetric documentation of the Red Gulch dinosaur tracksite (Wyoming, USA). In: Foss, S.E., Cavin, J.L., Brown, T., Kirkland, J.I. & Santucci, V. L. (eds), *Proceedings of the Eighth Conference on Fossil Resources*, pp. 101–103. Utah Geological Survey, Salt Lake City.

Bellian, J.A., Kerans, C. & Jennette, D.C. (2005) Digital outcrop models: applications of terrestrial scanning lidar technology in stratigraphic modeling. *Journal of Sedimentary Research*, **75 (2)**, 166–176.

Béthoux, O., McBride, J. & Maul, C. (2004) Surface laser scanning of fossil insect wings. *Palaeontology*, **47 (1)**, 13–19.

Bonnan, M.F. (2004) Morphometric analysis of humerus and femur shape in Morrison sauropods: implications for functional morphology and paleobiology. *Paleobiology*, **30 (3)**, 444–470.

Brasier, M.D. & Antcliffe, J.B. (2009) Evolutionary relationships within the Avalonian Ediacara biota: new insights from laser analysis. *Journal of the Geological Society*, **166 (2)**, 363–384.

Breithaupt, B.H. & Matthews, N.A. (2001) Preserving paleontological resources using photogrammetry and geographic information systems. In: Harmon, D. (ed), *Crossing Boundaries in Park Management: Proceedings of the 11th Conference on Research and Resource Management in Parks and on Public Lands*, pp. 62–70. The George Wright Society, Hancock.

Breithaupt, B.H., Matthews, N.A. & Noble, T.A. (2004) An integrated approach to three-dimensional data collection at dinosaur tracksites in the Rocky Mountain West. *Ichnos*, **11 (1–2)**, 11–26.

Einstein, A. (1916) Zur Quantentheorie der Strahlung. *Physikalische Gesellschaft Mitteilungen*, **16**, 47–62.

Einstein, A. (1917) Zur Quantentheorie der Strahlung. *Physikalische Zeitschrift*, **18**, 121–128.

Falkingham, P.L. (2012) Acquisition of high resolution three-dimensional models using free, open-source photogrammetric software. *Palaeontologia Electronica*, **15 (1)**, 1T.

Goswami, A. (2004) Cranial modularity across mammalia: morphometric analysis of phylogenetically and ecologically-related variation. *Journal of Vertebrate Paleontology*, **24 (Supplement to No. 3)**, 65A.

Green, D.J. & Alemseged, Z. (2012) *Australopithecus afarensis* scapular ontogeny, function, and the role of climbing in human evolution. *Science*, **338 (6106)**, 514–517.

Haring, A., Exner, U. & Harzhauser, M. (2009) Surveying a fossil oyster reef using terrestrial laser scanning. *Geophysical Research Abstracts*, **11**, 10714.

Jarvis, R.A. (1983) A perspective on range finding techniques for computer vision. *IEEE Transactions on Pattern Analysis and Machine Intelligence*, **PAMI-5 (2)**, 122–139.

Kearsley, A.T., Burchell, M.J., Graham, G.A., Hörz, F., Wozniakiewicz, P.A. & Cole, M.J. (2007) Cometary dust characteristics: comparison of stardust craters with laboratory impacts. *38th Lunar and Planetary Science Conference*, 1562.

Lukeneder, S. & Lukeneder, A. (2011) Methods in 3D modelling of Triassic ammonites from Turkey (Taurus, FWF P22109-B17). *Proceedings IAMG 2011*, pp. 496–505. Springer-Verlag, Salzburg.

Lyons, P.D., Rioux, M. & Patterson, R.T. (2000) Application of a three-dimensional color laser scanner to paleontology: an interactive model of a juvenile *Tylosaurus* sp. basisphenoid-basioccipital. *Palaeontologia Electronica*, **3 (2)**, 4A.

Maiman, T.H. (1960) Stimulated optical radiation in ruby. *Nature*, **187 (4736)**, 493–494.

Mallison, H., Hohloch, A. & Pfretzschner, H. (2009) Mechanical digitizing for paleontology – new and improved techniques. *Palaeontologia Electronica*, **12 (2)**, 4T.

Matthews, N.A., Noble, T.A. & Breithaupt, B.H. (2006) The application of photogrammetry, remote sensing and geographic information systems (GIS) to fossil resource management. *New Mexico Museum of Natural History and Science Bulletin*, **34**, 119–131.

Petti, F.M., Avanzi, M., Belvedere, M., et al. (2008) Digital 3D modelling of dinosaur footprints by photogrammetry and laser scanning techniques: integrated approach at the Coste dell'Anglone tracksite (Lower Jurassic, Southern Alps, Northern Italy). *Studi Trentini di Scienze Naturali, Acta Geologica*, **83**, 303–315.

Piazzesi, G. (1973) Photogrammetry with the scanning electron microscope. *Journal of Physics E: Scientific Instruments*, **6 (4)**, 392–396.

Platt, B.F., Hasiotis, S.T. & Hirmas, D.R. (2010) Use of low-cost multistripe laser triangulation (MLT) scanning technology for three-dimensional quantitative paleoichnological and neoichnological studies. *Journal of Sedimentary Research*, **80 (7)**, 590–610.

Remondino, F., Rizzi, A., Girardi, S., Petti, F.M. & Avanzini, M. (2010) 3D ichnology—recovering digital 3D models of dinosaur footprints. *The Photogrammetric Record*, **25 (131)**, 266–282.

Schuster, B.G. (1970) Detection of tropospheric and stratospheric aerosol layers by optical radar (lidar). *Journal of Geophysical Research*, **75 (15)**, 3123–3132.

Stoinski, S. (2011) From a skeleton to a 3D dinosaur. In: Elewa, A.M.T. (ed), *Computational Paleontology*, pp. 147–164. Springer, Berlin.

Wiedemann, A., Suthau, T. & Albertz, J. (1999) Photogrammetric survey of dinosaur skeletons. *Mitteilungen aus dem Museum für Naturkunde in Berlin, Geowissenschaftliche Reihe*, **2**, 113–119.

Wilhite, R. (2003) Digitizing large fossil skeletal elements for three-dimensional applications. *Palaeontologia Electronica*, **5 (2)**, 4A.

Zhang, G., Tsou, Y. & Rosenberger, A.L. (2000) Reconstruction of the *Homunculus* skull using a combined scanning and stereolithography process. *Rapid Prototyping Journal*, **6 (4)**, 267–275.

Further Reading/Resources

Bertolotti, M. (2004) *The History of the Laser*. Taylor & Francis, London.

Bogue, R. (2010) Three-dimensional measurements: a review of technologies and applications. *Sensor Reivew*, **30 (2)**, 102–106.

English Heritage. (2007) *3D Laser Scanning for Heritage*. English Heritage, Swindon.

Mallison, H. (2011) Digitizing methods for paleontology: applications, benefits and limitations. In: Elewa, A.M.T. (ed), *Computational Paleontology*, pp. 7–43. Springer, Berlin.

McGlone, J.C., Mikhail, E.M., Bethel, J.S., Mullen, R. (eds) (2004) *Manual of Photogrammetry*. American Society for Photogrammetry and Remote Sensing, Bethesda.

5

Digital Visualization

Abstract: Many possible workflows exist for converting data into virtual or physical models; those for tomographic data are more various and complex. Registered tomographic datasets are essential for most visualization techniques; their requirements are outlined, and methods for registration discussed. One alternative subsequent to this, vector surfacing, involves the tracing of outlines and their subsequent conversion to triangle meshes; it is now rarely used. Volume reconstruction, where pixels of tomograms are treated as three-dimensional voxels, is normally preferred. Volumes can be virtually 'prepared' to improve accuracy. They may be rendered directly, or through isosurfacing to generate a triangle mesh. Surface-based data may also be directly rendered or converted to a triangle mesh. Triangle meshes can be post-processed, and subsequently visualized using graphics hardware, ray tracing or three-dimensional printing. Selected visualization software packages are described, and the selection of data types and file formats for dissemination are discussed. Three visualization case-studies are presented.

5.1 Introduction

Chapters 2–4 have outlined the techniques required for three-dimensional data capture, through either tomography or surface-based approaches. These data are of limited use in their raw form. Virtual palaeontology, in the sense used herein, requires that the data are used to construct a digital three-dimensional model of the specimen, which then can be used as a basis for study or communication. Historically, tomographic data were either reconstructed using **physical modelling** techniques to produce cardboard or wax models (see Section 1.2), or in many cases no attempt was made to produce a three-dimensional visualization, data being presented simply as raw tomograms (e.g. Figure 2.2). While physical models have

Techniques for Virtual Palaeontology, First Edition. Mark D. Sutton, Imran A. Rahman and Russell J. Garwood.
© 2014 John Wiley & Sons, Ltd. Published 2014 by John Wiley & Sons, Ltd.

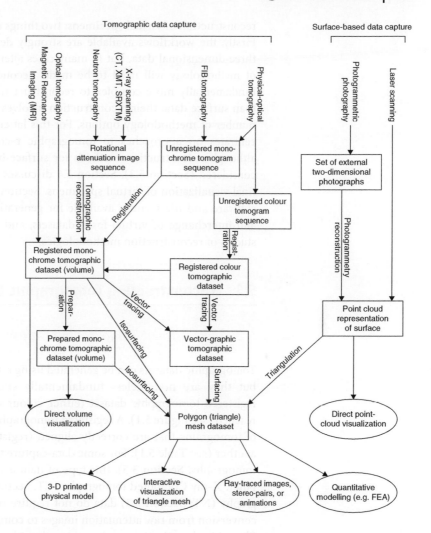

Figure 5.1 Common reconstruction workflows. Rectangles; dataset types generated during the reconstruction process. Ellipses; outputs (visualizations, models, etc.). Unbounded text at top; inputs. Modified from Sutton et al. (2012a, Fig. 2). Reproduced with permission of The Palaeontological Association.

some advantages, most obviously in their ease of visual inspection and suitability for display, they are in most ways inferior to digital visualizations. They are (for instance) more prone to reconstruction errors, difficult to copy, often too fragile to readily handle or move, impossible to dissect without damage, and time-consuming to construct. While digital-model production can also be labour-intensive when manual preparation work is undertaken (see Section 5.3.3.2), the minimum time-requirement to produce a model is normally much less than that of physical model making. Additionally, the benefits of physical models remain available when working digitally, as rapid-prototyping technologies that can three-dimensionally print a model are increasingly accessible (see Section 5.5.2.4).

This chapter describes and evaluates the software-based techniques required for digital reconstruction of raw data to create a final model. Figure 5.1 summarizes the most commonly encountered workflows in the reconstruction of virtual specimens; two things emerge from this diagram. Firstly, the workflows available are strongly dependent on the source of three-dimensional data, but in many cases alternatives exist, thus choices of methodology will need to be made. Secondly, tomographic data are fundamentally more complex to reconstruct from the user's perspective than surface data; their reconstruction involves more stages and a greater number of methodology options. For this latter reason, this chapter concentrates on approaches to tomographic reconstruction (Section 5.2), although reconstruction methods for surface-based data are also briefly considered (Section 5.3). Section 5.4 discusses the available methods for final visualization of virtual specimens. Section 5.5 discusses some of the software and file formats available for generation, manipulation, storage and interchange of virtual-fossil datasets, and Section 5.6 provides case studies of reconstruction methodologies.

5.2   Reconstructing Tomographic Data

5.2.1   Registered Tomographic Datasets

Tomographic datasets can be generated using a wide variety of techniques, but they are nonetheless fundamentally similar in nature once the registered tomographic dataset (either colour or monochrome) stage is reached (see Figure 5.1). A registered tomographic dataset is simply a series of tomograms that are correctly aligned (registered) with respect to one another (see Table 5.1). For some data-capture techniques (such as optical tomography, Section 3.5), this type of data is the raw input to the computer. X-ray Computed tomography (CT; Section 3.2) and neutron tomography (NT; Section 3.3) data do not require registration, but do require conversion from raw attenuation images to computed tomograms through filtered back projection or alternative algorithms; this process is normally considered to be part of the scanning methodology, and is hence described in Section 3.2.8. Datasets acquired with magnetic resonance imaging (MRI; Section 3.4) do not require registration either. In contrast, raw physical-optical (Section 2.2) and FIB (Section 2.3) datasets, which are directly captured in the form of tomograms, must normally be registered (Section 5.2.2). Registered datasets where the tomograms contain colour information can be kept in this format if the data are to be reconstructed via vector surfacing (Section 5.2.3). Where volume-based reconstruction approaches are to be employed, however (e.g. isosurfacing, Section 5.2.4.2), they are normally first converted to monochrome. This can be accomplished either using tools provided by reconstruction software, or using stand-alone image-editing packages.

Physical-optical datasets captured through manual tracing rather than photography are the exception to the general rule of similarity introduced at the start of this section. If a vector-surfacing methodology is to be used

Table 5.1 Assumptions about registered tomographic datasets made by all reconstruction methodologies.

Assumption	Notes
Pixel equivalence between tomograms	The vector offset in real space between all corresponding pixels (x, y) in tomograms n and $n+1$ must be perpendicular to the tomographic planes; in other words, images must be the same scale and resolution, must be aligned (registered) correctly with respect to each other, and must not contain distortions that cause regions of pixels to fail to correspond properly between tomograms. Achieving registration through simple rotation, translation and scaling of images is discussed in Section 5.2.2; registration is only normally required by physical-optical and FIB datasets. Distortions cannot easily be corrected for. These are primarily associated with physical-optical datasets, and can have a variety of causes (e.g. wrinkles on acetate peels, microscope lens-edge effects, tilting of photographed surfaces). They should be eliminated during data collection as far as is possible.
Parallel tomograms	All reconstruction methodologies assume that tomograms are parallel; we are not aware of any software that is able to correct for violations of this assumption, although if the angles between tomograms are known then compensatory retro-distortion is theoretically possible. Violations of this assumption are only likely to occur in physical-optical datasets.
Known spacing	Many reconstruction packages require that the spacing between tomograms is constant, but **all** require at least that it is known accurately for all tomogram pairs; merely knowing average spacing (for datasets where spacing is not guaranteed to be consistent) is not sufficient. Violations of this assumption may occur in either destructive or optical tomography unless steps are taken to avoid it during data capture.
Slice independence	All reconstruction methodologies assume that tomograms are fully independent, i.e. all data that appear on a tomogram are from a single plane; violation of this assumption technically means that the image is not a tomogram at all. Violations of this assumption can only be corrected for by manually identifying and removing offending data from each image. They are primarily associated with optical tomography, but can also occur in physical-optical tomography of translucent specimens; see also Section 2.2.3.4.

Note that further assumptions are made by volume-based approaches; these are detailed in Table 5.2.

(see Section 5.2.3), we recommend that the tracing stage is undertaken digitally from photographs (i.e. by tracing over images on screen after the production of a registered tomographic dataset – the 'vector tracing' steps in Figure 5.1). Tracing can, however, also be conducted manually, directly from the specimen or from acetate peels; many historical datasets are of this form. To reconstruct these tracings they must either be (a) scanned as images and subsequently treated as 'normal' physical-optical datasets, to be reconstructed using either vector surfacing or volume approaches, or (b) traced as vector-graphic objects using a digitizer (see Section 5.2.3). In the latter case, the raw input to the computer will be a vector-graphic tomographic dataset (see Figure 5.1); ensuring that this is registered may be problematic.

Reconstruction software of all types typically makes further assumptions about registered tomographic datasets (Table 5.1); where these assumptions are violated any final model is likely to be degraded. Some software may be able to partially correct for violations where they can be quantified; this is discussed in Table 5.1.

Once registered tomographic datasets have been generated, in many cases they can be cropped to size (also referred to as specifying a region of interest). **Cropping** involves (a) discarding any tomograms from the start or end of the dataset which contain no data of interest and (b) specifying a single rectangular region of interest to which all non-discarded tomograms can safely be cropped without excluding any data of interest. Cropping can, for instance, be used to remove fiduciary markings after registration. Note that the rectangular region must be consistently positioned and sized for all images to avoid violating the assumption of pixel equivalence between tomograms (see Table 5.1). If manual registration is required, it can obviously be skipped for tomograms which are to be cropped out of the final dataset. Cropping is not mandatory, but is a simple means to reduce the size of datasets and hence speed up computationally expensive processing during reconstruction.

5.2.2   Registration

Registration or alignment is the process of geometrically transforming individual tomograms to ensure that the requirement for pixel equivalence (see Table 5.1) is met throughout a dataset; in plain language, to ensure that the tomograms line up. Many tomographic techniques automatically register data, but a separate registration step is normally required for data generated by physical-optical tomography (Section 2.2) and FIB tomography (Section 2.3).

Registration comprises the translation (shifting, or horizontal displacement), rotation, and sometimes rescaling of tomograms, to ensure that correct alignment is achieved. Errors in pixel equivalence can cause significant degradation of final models, and hence the precision with which registration is carried out can be an important control on the quality of virtual reconstructions. Registration is greatly facilitated by the presence of fiduciary markings (see Section 2.2.3.3) in the tomograms. Where such markings are present, the correct position of a tomogram can be deduced simply by ensuring that the markings are in exactly the same position in each image. Note that partially sufficient markings (see Figure 2.4) may specify only certain transformations; a single edge, for instance, would allow the correct rotation and translation perpendicular to the edge to be deduced, but not the scale, or translation parallel to the edge. In the absence of fiduciary markings, the only guide to registration is the morphology of the specimen itself; this is fraught with difficulty as biological structures are complex and prone to unpredictable changes between tomograms. In our experience, accurate registration without fiduciary markings is not easy to achieve, and their all-too-frequent absence is a major obstacle to the successful reconstruction of historical datasets.

Registration can be either manual or automatic. In manual registration, the user is responsible for transforming each tomogram in turn, and judging by eye when suitably accurate registration has been achieved. Software tools for manual registration (e.g. SPIERSalign, Sutton et al. 2012a) provide means to compare adjacent images in the tomographic dataset so that registration can be judged, and/or overlay static markers on the images against which fiduciary structures can be positioned. **Automatic registration** involves algorithms which attempt to detect the position of fiduciary structures and/or the specimen itself, and use these to deduce the correct transformation required to achieve registration. While this approach is theoretically viable (see e.g. Watters and Grotzinger 2001), our experience with palaeontological physical-optical tomography datasets is that they are too noisy and inconsistent for it to be reliable; it may be more practical for FIB datasets. When automated alignment fails, it tends to fail badly; reliance on automated methods may lead to large errors in reconstructions. More useful is the concept of **automatic first-pass registration** (e.g. Sutton et al. 2012a), in which an automated registration attempt is made, with the expectation that the user will inspect and correct the registered dataset for errors using manual registration as a fall-back. This approach provides many of the time-saving advantages of automatic registration without relying on their success.

5.2.3 Vector Surfacing

There are two very different approaches to the production of reconstructions from registered tomographic datasets; the first of these, here termed vector surfacing, uses a vector-graphics model. The majority of early attempts at reconstruction from physical-optical datasets (e.g. Chapman 1989; Herbert 1999) used this approach, which is equivalent to the 'surfaces' approach of Sutton et al. (2001); our new term is less easy to confuse with isosurfacing (Section 5.2.4.2). Vector surfacing involves the manual or semi-automatic identification of structures of interest in each tomogram (Figure 5.2a, b) in the form of vector-graphic objects, normally closed loops defined as spline curves (Figure 5.2c). These two-dimensional splines are then stacked in virtual three-dimensional space, and used to generate a mathematically defined three-dimensional surface, typically in the form of a triangle mesh (Figure 5.2d). As noted in Section 5.2.1, we recommend that these spline objects are created on screen by tracing each tomogram of a (photographic) registered tomographic dataset; this approach allows standard registration tools to be used. Alternatively, they may be created using a digitizer (graphics tablet) device to directly generate vector-graphics objects.

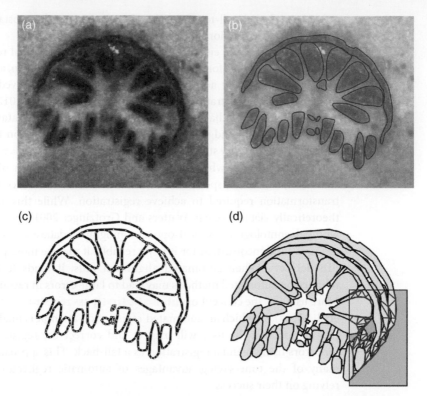

Figure 5.2 Vector surfacing reconstruction. (a) Photographic tomogram. (b) Photographic tomogram with tracing overlaid. (c) Traced structures converted into vector-graphic objects (spline curves – nodes shown). (d) Spline curves from multiple tomograms algorithmically surfaced, here with a triangle mesh (blue and red). Modified from Sutton et al. (2001, Fig. 4). Reproduced with permission of The Palaeontological Association.

Vector surfacing has some advantages over volume-based reconstruction (Section 5.2.4). It performs better where spacing between tomograms is relatively high, and is relatively simple to adapt to datasets where tomographic spacing is inconsistent. While it cannot recreate information lost between widely spaced tomograms, smoother interpolations can be generated. It also typically produces triangle meshes that are more compact (i.e. use fewer triangles) than those produced by isosurfacing methods (Section 5.2.4.2). Finally, it is not sensitive to variations in lighting or other optical conditions between tomograms; as long as a tracing can be made, the approach is viable. The requirement for a tracing stage can be viewed as either an advantage or a disadvantage. A user is required to make active decisions as to exactly where to place splines; this is time-consuming and can be viewed as unnecessarily subjective. It does, however, ensure that all positioning of surfaces is underpinned by considered interpretation of data rather than relying on automation; a human eye is less likely to be fooled by imperfections or other spurious data in tomograms. In general, vector surfacing performs well for simple objects (e.g. individual vertebrate bones), but not for complex objects that merge and split between tomograms, as correspondence between spline objects from tomogram to tomogram can be problematic to automatically determine during surfacing. In some software packages, such connectivity must be manually specified; others may use automated but error-prone approaches. This weakness is the corollary of the advantage of smoother interpolation, as interpolation requires an understanding of what must be interpolated between.

Automated and semi-automated approaches to the generation of spline objects have also been used. Herbert (1999) described an algorithm that used a model of brachiopod anatomy to identify structures such as the dorsal and ventral valves, and to 'grow' spline objects using this model in a fault-tolerant way through a set of tomograms. This approach, while at least partially successful, is reliant on a degree of prior knowledge of the specimen and is hence not applicable to generalized objects; it has not seen further application. Maloof et al. (2010) described an innovative vector-surfacing approach where the auto-trace facility of commercial vector-graphics software was used to identify structures; this is detailed in Section 5.6.3.

Vector surfacing has become a rare approach to reconstruction in recent years. For datasets where tomogram frequency is high (e.g. those from scanning methodologies such as CT or NT), it has no telling advantages over isosurface-based volume reconstruction other than the production of models with a lower triangle count, a consideration which is at least partially offset by the power of modern computers. Furthermore, secondarily smoothed isosurfaces approach vector-surfaced models in terms of the quality of interpolation between tomograms, and volume datasets can also be interpreted on a tomogram-by-tomogram basis if so desired (see Section 5.2.4.3). Vector-surfacing software packages are also less generally available. For these reasons, we recommend this approach only for physical-optical datasets where (a) tomogram spacing is high (more than four times the pixel-spacing within tomograms), and (b) objects are relatively simple. No reconstruction methodology will adequately reconstruct complex specimens where tomogram spacing is too high for either a human or an algorithm to determine connectivity between objects.

5.2.4 Volume Reconstructions

5.2.4.1 Concepts and Assessment

Volume reconstruction is an alternative to vector surfacing; the approach has been used in the majority of virtual palaeontology studies in this century. Rather than modelling a fossil with vector-graphics objects, a raster-based (pixel-based) approach is taken. The dataset is treated as a volume comprising cubic or sub-cubic voxels. Voxels are simply three-dimensional pixels; in the same way that a two-dimensional raster image comprises a rectangular grid of equally sized pixels, a volume comprises a three-dimensional grid of equally sized voxels. Voxels, like pixels, each

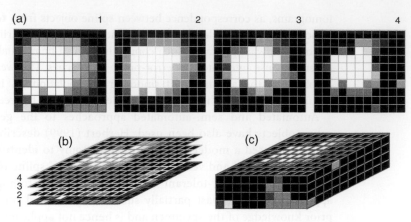

Figure 5.3 Volume representation. (a) A monochrome registered tomographic dataset consisting of four low-resolution tomograms 1–4. (b) Tomograms 1–4 conceptually stacked in virtual space. (c) A volume based on the tomograms, where pixel values (levels of grey) are treated as voxels with the same centre. This volume is isotropic – voxels are equidimensional (i.e. spacing between tomograms is equal to pixel spacing within tomograms).

represent a single measurement of colour (or brightness for monochrome datasets). A registered tomographic dataset of raster images with regular tomogram spacing does not need conversion into a volume; only the conceptual shift of considering pixels as voxels is required (Figure 5.3).

A volume can be visualized (as a three-dimensional reconstruction) in one of two ways – by direct volume rendering (Section 5.4.3) or by the extraction of an isosurface (Section 5.2.4.2). Both can be accomplished automatically and quickly on modern computers. Volumes can also be rendered in two dimensions by 'reslicing', providing new virtual sections in any desired plane, at the risk of introducing resampling artefacts. The key advantage of volume-based reconstructions is their speed and simplicity from a user's perspective; visualizations can be prepared quickly and with minimal intervention, as there is no required interpretation step. While interpretation can be desirable (see Section 5.2.3), the facility to perform it is available if required; see Section 5.2.4.3. Reconstructions using either volume visualization approach are at least as visually satisfactory as those prepared by vector surfacing, but do apply further strictures on the underlying registered tomographic dataset; these are detailed in Table 5.2. For 'scanner' datasets (CT, NT, MRI), however, none of these restrictions are problematic, and although they may complicate reconstruction of other datasets, they rarely present insurmountable obstacles. Volume-based reconstruction is therefore the method of choice for the vast majority of modern tomographic datasets, palaeontological or otherwise.

5.2.4.2 Isosurfaces
The most commonly used means to visualize a volume involves the calculation of an isosurface: a three-dimensional surface that connects all points

Table 5.2 Assumptions about registered tomographic datasets made by volume-based reconstruction approaches.

Assumption	Notes
Consistently spaced (sub)-isotropic data	Voxel spacing in volumes should be consistent; violations are most commonly the result of inconsistent tomogram spacing in physical-optical datasets. Some software is able to compensate for such violations using approaches such as differential stretching of an isosurface model (e.g. SPIERS, Sutton et al. 2012a), but the most satisfactory reconstructions are achieved where spacing is consistent. A volume should also ideally be isotropic, i.e. pixel spacing within a tomogram should be the same as pixel spacing between tomograms (Figure 5.3c). 'Scanner' data (CT,NT,MRI) are normally isotropic, but data from other forms of tomography are usually not. Some reconstruction software (e.g. SPIERS, Sutton et al. 2012a) can compensate for divergences from isotropy, but divergence by a factor of more than 3 or 4 is not recommended for either isosurface or direct volume rendering. Where this divergence is caused by relatively high tomogram spacing, it can be corrected by downsampling (='binning') tomograms in a dataset (i.e. reducing their resolution), although this may reduce the information content of the tomograms.
Consistency of lighting	Treating data as a volume requires that all pixels (to be treated as voxels) are captured under consistent conditions. Variations in brightness, contrast or colour balance between tomograms violate this requirement, and can severely degrade reconstructions. Ideally, this should be controlled for during data collection, but post-processing in image-editing software to achieve consistent lighting conditions may be necessary in some cases.
Monochrome data	Isosurfacing and many direct volume-rendering techniques require a monochrome volume, where each voxel is represented by a single value. While it is technically straightforward to convert colour datasets to monochrome, this may involve discarding useful information.
Unambiguous segmentation (isosurfaces only)	The isosurface-based reconstruction approach assumes that sharp boundaries exist between fossil and matrix, or between different types of fossilized tissue. Material where fossil grades into matrix, or one tissue type grades into another, is not suitable for reconstruction using this methodology, unless the arbitrary selection of a boundary level in this gradation is acceptable.

of a constant (user-determined) intensity within the volume. Thresholding of individual tomograms (setting to black pixels below the defined intensity level, and to white those above – or vice versa where fossils are darker than matrix) is a useful way to conceptualize the isosurface prior to reconstruction, and software that reconstructs using this methodology normally provides users with this view of the data (Figure 5.4c, e). Isosurfaces are usually generated using the **marching cubes** algorithm (Lorensen and Cline 1987), which produces a triangle-mesh dataset defining the isosurface (or surfaces; there is no requirement for all points to connect into a single surface). Once generated, isosurfaces can be visualized using the approaches described for triangle meshes, which bring many practical advantages (Section 5.4.2).

The marching cubes algorithm is robust and capable of generating isosurfaces for any dataset. Its primary drawback is the large number of triangles that it generates, especially for 'noisy' data with a large number of isolated pixels. A noisy $1024 \times 1024 \times 1024$ volume (a typical resolution for a modern

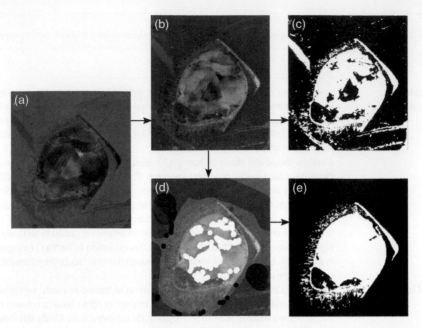

Figure 5.4 Thresholding and preparation of a tomogram (from holotype of *Kulindroplax perissokomos*, Sutton et al. 2012b). (a) Colour tomogram. (b) Monochrome tomogram derived from (a) – inverted so originally dark pixels are light. (c) Thresholded version of (b); note that while portions of the fossil are well picked out, some material that is fossil is black (e.g. the light crystals in the centre), and some material that is not fossil is white (e.g. the red band in the matrix below the lettering). (d) Prepared version of (b) – some regions have been darkened, others lightened; in some cases pure black and white have been applied. (e) Thresholded version of (d); note that the issues documented in (c) have been fixed.

XMT scanner) can easily result in a 200 million triangle model, beyond the capabilities of most visualization software or hardware. Mesh-simplification algorithms (Section 5.4.2.1) can help mitigate this problem, but are themselves memory-intensive and can also be overwhelmed by a mesh of this scale. Tackling this problem prior to isosurface reconstruction is often the only practical means to control mesh size. This can be accomplished either by removal of noise by virtual preparation (Section 5.2.4.3), downsampling or 'binning' of the volume (e.g. converting a 1024 × 1024 × 1024 volume to 512 × 512 × 512), or both.

5.2.4.3 Virtual Preparation of Volumes

Volume data can be reconstructed 'raw', but in most cases reconstructions can be greatly improved by preparation work; this is very much the virtual equivalent of the manual preparation work traditionally employed to physically expose specimens (see e.g. Whybrow and Lindsay 1990). Volume preparation involves either modifying the values of the voxels by brightening or darkening them (see e.g. Sutton et al. 2001), applying **masks** (see e.g. Abel et al. 2012; Sutton et al. 2012a), or refining visualization rules in an attempt to improve the discrimination of the structures of interest.

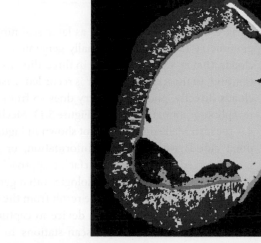

Figure 5.5 Masked and thresholded tomogram of the holotype of *Kulindroplax perissokomos* Sutton et al. (2012b). Masks have been applied manually to identify individual structures and regions; these are displayed as regions of colour. The thresholded tomogram under the masks is that of Figure 5.4e.

Depending on the capabilities of the software used, volume preparation can be carried out either in two dimensions, on a tomogram-by-tomogram basis, or in three dimensions. The latter approach is generally faster, but allows for less precision and inspection of work, and does not normally allow modification of voxel values.

Voxel modification is normally used in conjunction with isosurface-based visualization (Section 5.2.4.2), in which voxels are thresholded. In any real dataset, no single threshold value can satisfactorily segment all pixels correctly into 'specimen' and 'not specimen' (see Figure 5.4c). Voxel modification can be used to manually fix such errors by darkening or blackening structures that are not desired in the reconstruction, and brightening structures that are desired but are not reaching the set threshold value (see Figure 5.4d, e). Voxel modification, especially when carried out on a tomogram-by-tomogram basis, allows careful and precise determination of exactly what structures will or will not be included in a visualization. It is a time-consuming task requiring considerable skill, especially when the dataset is noisy and contains imperfections, but can greatly increase the fidelity and hence scientific value of any resulting visualization.

Masks (= **labels** or **segments** in some software) represent data overlays on the volume that flag individual voxels as belonging to one or more objects; they are typically represented by colour overlays (Figure 5.5). Masks are used by reconstruction software to split models into arbitrarily shaped areas; in any visualization the user can choose how different masked areas display, or exclude them entirely. Applications of masking include the removal of unwanted objects that may obscure the specimen, the flagging of biologically significant structures with colour, the separation of composite fossils, the performance of virtual dissection through selective removal of objects, or the translucent rendering of selected objects.

5.3   Reconstructing Surface Data

Surface-based techniques such as laser scanning (Section 4.2) and photogrammetry (Section 4.3) normally generate datasets in the form of point clouds, that is, a series of points in three-dimensional space in which a position and, in most cases, a colour is recorded. Laser scanning generates point clouds directly; photogrammetry does so from algorithmic analysis of the raw photogrammetric dataset (Figure 5.1). Mechanical digitization (Section 4.4, see also Mallison 2011) is not shown in Figure 5.1; it can generate either point clouds without colour information, or spline curves that can be reconstructed using a vector-surfacing approach (Section 5.2.3).

Surface data-capture methodologies often generate several discrete datasets for a single specimen. These result from the need to move the specimen with respect to the acquisition device to capture all surfaces (e.g. Section 4.4), or the use of multiple scan-stations to ensure full coverage (e.g. Section 4.2.3.2). The process of fusing these datasets into one is normally termed registration, but should not be confused with **tomographic registration** (Section 5.2.2). Surface-data registration is achieved through the identification of identical structures in the separate datasets, enabling their relative positions to be determined. It is normally carried out automatically or semi-automatically by software associated with the acquisition device.

Point clouds can be directly visualized (see Section 5.4.3), or can be surfaced through triangulation algorithms to produce a triangle mesh. The latter approach allows conversion of the model into a similar format to that produced by other virtual palaeontological workflows and increases the range of visualization options. It is, however, not computationally straightforward; while effective algorithms exist (see e.g. Marton et al. 2009, Salman et al. 2010), they are still the subject of active research, and not always integrated into reconstruction software. **Direct point-cloud visualization** may hence be the simplest option for surface data.

5.4   Visualization Methodologies

5.4.1   Introduction

Sections 5.2 and 5.3 detail the means by which visualization-ready datasets (triangle meshes, volumes, or point clouds) can be produced from raw tomographic or surface data. As Figure 5.1 makes clear, visualization of triangle meshes is the most widely applicable means by which this can be accomplished, and provides the broadest range of visualization options. This is also the approach responsible for the majority of published palaeontological virtual specimens, and it is hence treated here in the most detail (Section 5.4.2). Sections 5.4.3 and 5.4.4 briefly discuss two alternative techniques.

Figure 5.6 The effects of mesh simplification and smoothing algorithms on *Offacolus kingi* isosurface mesh (OUMNH C.29558, figured in Sutton et al. 2002, Fig. 2). Scale bar is 1 mm. Ktr = 1000 triangles. Mesh algorithms performed using SPIERS 2.15 (Sutton et al. 2012). (a) Unprocessed isosurface, 1156 Ktr. (b) Isosurface simplified using decimation algorithm, 689 Ktr. (c) Isosurface simplified using quadric error metrics algorithm, 404 Ktr. (d) Isosurface simplified using quadric error metrics algorithm, 296 Ktr; note visible artefacts towards the top of the specimen. (e) Unsimplified isosurface weakly smoothed using iterative windowed sinc algorithm (see Bade et al. 2006, p. 12). (f) Unsimplified isosurface more strongly smoothed using iterative windowed sinc algorithm; note the damage to the fine hairs at the top of the specimen.

5.4.2   Visualizing Triangle Meshes

5.4.2.1   Mesh Processing

Triangle meshes, particularly those produced by isosurfacing (Section 5.2.4.2) and point-cloud surfacing (Section 5.3), can be very large in terms of triangle count; we have frequently dealt with meshes comprising over 100 million triangles. These may be impractical to visualize on anything but high-end computers (i.e. those with a large quantity of RAM and a very powerful graphics subsystem). Many algorithms exist that can reduce the triangle count of a mesh while minimizing damage to the overall surface topography; in some cases, these are an acceptable means of reducing the triangle count to manageable levels. The most commonly encountered are normally referred to as **decimation** algorithms (Schroeder et al. 1992) and **quadric error metric algorithms** (Garland and Heckbert 1997, 1998). The latter normally produce better results (i.e. better preservation of surface form for any given target triangle count), but are more memory-intensive, and may themselves require computers with a large amount of RAM to run successfully. Depending on the algorithm implementation, attempts to reduce triangle count beyond certain limits may simply fail (i.e. the triangle count may not reduce below a certain level), or result in unacceptable artefacts, in particular degrading detail in structures such as hairs, spines, etc. Figure 5.6 shows the effects of successful (Figure 5.6b, c) and unsuccessful (Figure 5.6d) mesh simplification.

Triangle meshes produced by isosurfacing may also be 'blocky' in detail, reflecting roughness from the underlying voxel grain of the volume. Algorithms that subtly shift mesh points to provide a smoothing effect (see Bade et al. 2006) can be used to combat this problem, but should be applied with care (e.g. Figure 5.6e); if too strong a smoothing effect is employed, morphological information may be lost (e.g. Figure 5.6f).

Simple mesh-processing algorithms can also be used to perform **island removal**, identifying all connected areas and removing any below a certain size. This can be a quick way to remove noise from a dataset, as typically the largest connected region will be the specimen itself, and small disconnected regions will be spurious. Care should be taken, however, to ensure that no disconnected parts of the specimen itself are inadvertently removed in this way.

Some triangle meshes, particularly those generated by surface-based techniques, may contain small holes. These may be problematic for visualization, and are more commonly so for other applications of virtual models (see Chapter 6). Mesh-processing algorithms exist to fill such holes, and are thus of occasional use in virtual-fossil reconstruction workflows.

5.4.2.2 Hardware-Accelerated Triangle-Mesh Rendering

One important reason to use triangle-mesh models as a means of representing virtual specimens is the wide availability of cheap and immensely powerful **triangle-mesh rendering** hardware. While **polygon-mesh** rendering (of which triangle-mesh rendering is the most important form) has many worthy applications in science, engineering and medicine, it is its use in the video games industry that has been the most important driver for the technologies involved. As a result, even the lowest-specification modern computers possess significant capabilities for hardware-accelerated rendering of data of this type. Commodity 'graphics cards' are, in essence, highly optimized polygon-rendering devices; workstation-class machines possess even more capable equivalents. Standardized graphics libraries such as OpenGL exist to provide straightforward access to this hardware. Direct on-screen rendering of highly detailed models is thus possible on even modest computers, and this facility is provided by most if not all reconstruction software.

While hardware-accelerated rendering can be used to provide the researcher with a few pre-rendered static images of a specimen, or a video of a pre-defined rotation, the real power of this approach lies in interactive visualization. Here, the fast rendering speed is used to implement an interactive system where the user can manipulate a virtual model, rotating, zooming, and altering visibility of discrete elements at will. This approach provides an extremely powerful means of visualizing and exploring data. This power can be further augmented by stereoscopic (three-dimensional) viewing of the specimen, either using **anaglyph stereo** (viewed with red/green or red/blue filtered glasses) or any of the more technologically refined three-dimensional stereoscopic viewing systems now available. The authors have found that interaction with a virtual dataset in this manner is a powerful way to explore the morphological information that it contains.

Despite the power of the available hardware, the large size of some virtual-fossil datasets can make interactive rendering difficult in practice,

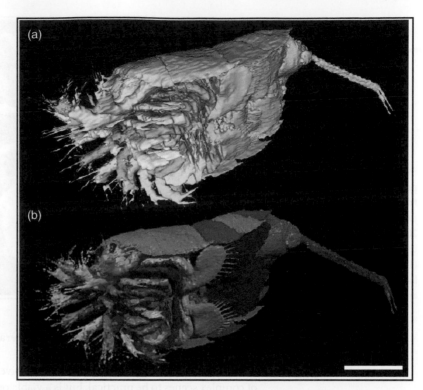

Figure 5.7 Visualizations of *Offacolus kingi* (OUMNH C.29558, figured in Sutton et al. 2002, Fig. 2); scale bar is 1 mm. (a) OpenGL rendering. (b) Ray-traced rendering.

especially on low-powered computers. While rendering performance is affected by most aspects of system specification, the quantity of dedicated graphics RAM is typically the most important limiting factor, performance normally dropping rapidly once this memory limit is reached. The exact model size to which this limit corresponds will depend on details of hardware and also on the efficiency of the software in use. Our experience with the SPIERSview rendering system (Sutton et al. 2012a) is that models of 20–30 million triangles in size will render at 20–30 frames per second on high-end laptop or mid-range desktop computers with 1–2 Gb of graphics RAM, but will render at much less than one frame per second on low-end laptops which lack powerful graphics subsytems. For satisfactory interactive rendering, our experience is that a frame rate of at least three frames per second is required.

5.4.2.3 Ray Tracing

Ray tracing is a rendering approach based on a detailed implementation of the physics of optics; it is generally considered to be amongst the most photo-realistic means of visualization available (see e.g. Glassner 1989). In particular, it handles lighting in a way that provides realistic reflections and shadows. The method can be used to visualize any mathematically defined surface, which includes triangle meshes, but not point clouds or directly rendered volumes. Figure 5.7 shows a ray-traced rendering

Figure 5.8 Three-dimensional printed model of the trigonotarbid arachnid *Eophrynus prestvicii* (see Garwood et al. 2009), produced from CT data using laser sintering technology (DTM2500+ Sinterstation). Scale bar is 10 mm (model is approximately 4.7 × natural scale).

placed next to an equivalent rendering generated using the techniques of Section 5.4.2.2.

Ray tracing is too computationally expensive for interactive rendering of complex scenes to be practical, but is a viable option for the production of high-quality static images or stereo pairs, or of pre-rendered animations (see e.g. Siveter, Derek J., et al. 2004). The approach has not been widely adopted for palaeontological visualization, probably as reconstruction software suites do not normally implement ray-tracing capabilities, and the importation of models into dedicated ray-tracing packages (such as the open-source package Blender, Section 5.6.1.5) is required. While in some cases the addition of realistic shadows may obfuscate morphology, for most specimens we consider ray-traced images to be the optimal means of presenting non-interactive visualizations, and recommend that ray tracing is employed where possible.

5.4.2.4 Three-Dimensional Printing

Three-dimensional printing encompasses a variety of approaches for automatically making a three-dimensional solid object from a digital model. These processes, also known as rapid prototyping or additive manufacturing, have been developed primarily for manufacturing and engineering; several alternative technologies exist including stereolithography and laser sintering (see Gibson et al. 2010 for a full review). Triangle-mesh models of virtual palaeontological specimens are readily printable using these technologies (see Figure 5.8), in a range of colours, materials and sizes. Drawbacks exist: firstly, reduction in triangle count is often required to conform to the capabilities of the printer; this must be achieved either through downsampling ('binning') of data prior to reconstruction, or mesh simplification, and

Figure 5.9 Volume ray-casting image of a crocodile mummy. (CT data; Phoebe A. Hearst Museum of Anthropology, UC Berkeley).

either can reduce fidelity of the model. Additionally, models can be fragile, often fail to retain subtly connected structures that lack the physical strength to hold together as one object, are not cost-free to produce, and unlike virtual models cannot be easily copied, disseminated or dissected. Nonetheless, physical models are appealing in providing a physical reconstruction which can be directly inspected. They also have potential as tools for model **validation**, especially in **computational fluid dynamics** (see Section 6.4.4). Perhaps their widest use could be in the communication of results to academic and, increasingly, non-specialist audiences (see e.g. Rahman et al. 2012).

5.4.3 Direct Volume Rendering

A variety of algorithms exist for the direct visualization of volume data, obviating the need for calculation of an isosurface; see, for example, Lichtenbelt et al. (1998) for a review. These approaches vary in detail and in computational efficiency, but all share the concept of projecting a volume directly onto a two-dimensional image. The most easily understood and most widely used variant, **volume ray-casting**, can be conceptualized as a virtual 'X-ray' of the volume, in which the colour and/or brightness of each pixel in the resultant image depends on the properties of all voxels through which a virtual light-ray has passed. Other approaches exist, but are beyond the scope of this book.

Volume ray-casting copes well with specimens in which gradations are present, as no arbitrary thresholding is required (c.f. Section 5.2.4.2). It also automatically implements translucency of structures, which may be desirable, and is capable of preserving any colour information in the original volumes (not shown in Figure 5.1 for clarity). Results can be visually appealing, rivalling those achievable from triangle-mesh visualizations (Figure 5.9). The approach has nonetheless not been widely used for palaeontological

Figure 5.10 Direct point-cloud visualization of fossil tree-root system, Manchester Museum. Dataset is photogrammetrically generated, and consists of 841,059 points. Source: Falkingham (2012, Fig. 8). Reproduced with permission of The Palaeontological Association.

datasets (though see Schiffbauer and Xiao 2009; Albani et al. 2010); while it may be desirable for certain types of material (see the preceding), in general, there are no strong incentives to prefer it over triangle-mesh-based visualization. **Volume rendering** is often associated with relatively slow rendering times, more in line with ray tracing (Section 5.4.2.3) than **hardware-accelerated triangle-mesh rendering** (Section 5.4.2.2). Hardware acceleration using commodity graphics cards has been widely implemented in recent years (see e.g. Stegmaier et al. 2005), resulting in substantial improvements in rendering times, but nonetheless real-time interactive volume rendering remains more demanding in hardware terms than interactive rendering using triangle meshes. It also requires the presence of the full volume dataset, which can hamper data-sharing. Software availability is a further limiting factor on uptake; while capable software packages exist, they are less diverse than those available for triangle-mesh rendering, and may not always be accessible to researchers.

A terminological confusion exists with respect to volume rendering, namely that some workers use the term to refer to any form of rendering derived from volume data, including the calculation of an isosurface. Many (perhaps most) examples of 'volume rendering' in the palaeontological literature are thus not direct volume rendering, but renderings of a triangle-mesh isosurface; Penny et al. (2007) provides an example of this.

5.4.4 Direct Point-Cloud Rendering

The direct rendering of point clouds is a relatively straightforward process, involving the projection of points from three into two dimensions, drawing each as a small coloured square or circle in the appropriate position. Figure 5.10

provides a palaeontological example. **Direct point-cloud rendering** is normally less visually appealing than triangle-mesh-based rendering and often slower, as it is less amenable to hardware acceleration (though see e.g. Wimmer and Scheiblauer 2006). However (as noted in Section 5.3), it avoids the non-trivial triangulation step and is hence simpler to achieve. For dense point clouds it may be an adequate means to visualize surface data, and where points are sparse it has the benefit of making this sparsity clear to the viewer.

5.5 Software and Formats

5.5.1 Reconstruction and Visualization Software

The computer reconstruction of a virtual fossil requires software. One approach is to use non-specialized software for different parts of the reconstruction process, where necessary using custom-written scripts or programs to improve efficiency or convert data between formats as part of the reconstruction pipeline. Examples of this approach include Sutton et al. (2001) and Maloof et al. (2010). These solutions can be effective, but are typically specific to particular workflows, which are, in turn, normally specific to particular types of data. These methods hence lack versatility, and need re-designing if they are to be applied to new forms of data. To do so normally requires programmer-level IT skills. For this reason, most virtual-fossil reconstructions published in recent years have used pre-existing reconstruction packages to handle as much of the reconstruction workflow as possible.

A range of software packages for three-dimensional reconstruction exist. These vary enormously in price, capability, ease of use, and in the methodologies implemented. Software packages for the reconstruction of X-ray CT data, in particular, are plentiful, and are easily adapted to other tomographic techniques that produce isotropic monochrome volumes (e.g. neutron tomography). A full review of all commercially available software is beyond the scope of this book. Other than to note that many recent publications of CT-datasets have made use of the commercial packages Amira and Avizo (www.vsg3d.com – see e.g. Gai et al. 2012; Young et al. 2012) and VGStudio Max (www.volumegraphics.com/en – see e.g. Abel et al. 2012), we restrict our discussion here to two selected free software packages for tomographic reconstruction and visualization (SPIERS and Drishti), and three that are useful for parts of reconstruction workflows (ImageJ, Meshlab and Blender).

5.5.1.1 SPIERS

SPIERS (Serial Palaeontological Image Editing and Rendering System, www.spiers-software.org; Sutton et al. 2012a) is a free general-purpose tomographic reconstruction suite specifically designed for the reconstruction of palaeontological specimens. It is available for both Windows and OSX, and is provided with detailed manuals written with palaeon-tologists in mind. It uses an isosurface-based reconstruction methodology; while it incorporates vector-based tracing tools, it cannot perform vector surfacing *per se*, and it has no support for direct volume rendering. SPIERS was initially developed to reconstruct physical-optical datasets (see Sutton et al. 2012a and references therein); it hence incorporates a registration tool, and has facilities to handle non-isotropic data, inconsistently spaced tomograms, and colour datasets. While developed for physical-optical data, the package has also been used for CT-type datasets (see e.g. Selden et al. 2008; Garwood et al. 2009; Rahman and Clausen 2009), and is theoretically applicable to any tomographic data. SPIERS does not include import tools; it requires data in the form of image sequences, and conversion to this form from other formats (e.g. DICOM from CT scanners) is left to the user to implement in other software (e.g. ImageJ, Section 5.5.1.3).

SPIERS provides simple but powerful tools for virtual preparation of volumes (see Section 5.2.4.3), through masking, segmentation (SPIERS uses this term to mean the differentiation of material on the basis of colour or brightness), and voxel modification. While some volume-based tools are included, it is primarily a 'slice-by-slice' editor, whose design philosophy is that users undertaking virtual preparation will value slow-and-careful working up of their data to extract maximum information.

SPIERS also includes a stand-alone hardware-accelerated interactive viewer with mesh pre-processing capabilities (SPIERSview), notable for its lightweight system requirements that allow large models to be viewed on modest computers. As well as providing rendering and model export capabilities for isosurface models from the SPIERS package itself, SPIERSview is capable of viewing any triangle-mesh dataset in **VAXML** format (see Section 5.5.2), including those generated from surface-based methodologies. SPIERSview only includes a rudimentary animation system, but is capable of exporting models to third-party software (e.g. Blender, Section 5.5.1.5) for the production of ray-traced animations.

5.5.1.2 Drishti

Drishti (anusf.anu.edu.au/Vizlab/drishti/) is a free and open-source tomography package with an emphasis on visualization tools; see, for example, Long et al. (2008) for a palaeontogical application. It is available for Windows, OSX and Linux, and has an active support community. Drishti provides both isosurface-based triangle-mesh rendering and direct volume rendering, and its animation and rendering tools are sophisticated. It also provides tools for three-dimensionally masking and region identification, but provides only limited support for detailed virtual preparation of volume data. It does, however, have an extensive range of import options, streamlining its use on many datasets, and its flexible and high-quality rendering and animation systems may mean that no export to a final rendering package is required.

Drishti is designed for CT-type isotropic monochrome volumes. Its design philosophy is radically different from that of SPIERS, emphasizing the fast production of high-quality visualizations from relatively 'clean' data, rather than the slow preparation of noisy data into models with maximal information content. It is thus not well-suited to the noisy and non-isotropic datasets generated by some tomographic methodologies (e.g. physical optical, focused ion beam, some optical tomography), or indeed to noisy CT data from 'difficult' specimens. It does, however, provide an effective all-in-one solution for the rapid visualization of high-quality tomographic datasets.

5.5.1.3 ImageJ

ImageJ (rsb.info.nih.gov/ij/index.html) is a free and open-source general-purpose image manipulation tool that can work on 'stacks' of images, such as tomogram sequences. Written in platform-agnostic Java, it is available for all common platforms including Windows, OSX and Linux. It is well-documented and widely used, as is the related Fiji (fiji.sc), which comes packaged with a larger range of plugins. ImageJ is not primarily a three-dimensional reconstruction tool or an image editor, but excels at automatic operations such as contrast alterations, cropping of volumes, the application of filters such as noise reduction or sharpening, and the conversion of file formats (including the conversion of DICOM files into image sequences). It is scriptable and highly extensible; a large range of free third-party plugins are available, including three-dimensional visualization tools. Most virtual palaeontology workflows are hence theoretically implementable in ImageJ, although efficiency in terms of rendering performance and hardware requirements is, in our experience, generally less than that of dedicated packages. ImageJ is aimed at relatively IT-literate users, and implementation of efficient workflows will normally involve the use of its built-in scripting facility. While many researchers may find ImageJ useful for file conversion and batch data processing, we do not recommend it as a platform for full three-dimensional reconstruction except to researchers with programmer-level skills.

5.5.1.4 Meshlab

Meshlab (meshlab.sourceforge.net) is a free and open-source general-purpose triangle-mesh processing and visualization tool. It is available for Windows, OSX and Linux, is easy to use, and is well-supported and documented. Cut-down versions of Meshlab are also available for mobile/tablet operating systems (iOS and Android), and can be used to view data on more portable devices. Meshlab supports a wide variety of mesh file types, and can be used as a format converter. It is also a capable interactive viewer for both triangle-mesh and point-cloud datasets, and includes triangulation tools to convert the latter into the former (see e.g. Falkingham 2012). Meshlab provides reslicing tools and extensive triangle-mesh processing facilities, including all those described in Section 5.4.2.1. It is a powerful tool for the manipulation of triangle-mesh-based datasets, and best used to augment the capabilities of packages such as SPIERS and Drishti. Note that its interactive viewer lacks the performance of, for example, SPIERSview (see Section 5.5.1.1) for large datasets on modest computers.

5.5.1.5 Blender

Blender (www.blender.org) is a free and open-source three-dimensional animation and ray-tracing (Section 5.4.2.3) package, available for Windows, OSX and Linux. It is widely used and well-documented, and has been used to produce animations and static images for palaeontological publications (e.g. Garwood et al. 2009; Sutton et al. 2012b; Zamora et al. 2012). Blender is not a simple package to use, having a somewhat quirky interface and requiring a new user to undertake substantial conceptual learning. Once mastered, however, it provides powerful ray-traced visualization capabilities rivalling those of commercial software. Note that Blender does provide interactive three-dimensional visualization, but we do not recommend its use in this mode; it is primarily a ray-tracing package, and we recommend that, in a virtual-fossil context, it is used primarily for the creation of ray-traced static images, stereo pairs and animations.

5.5.2 *Data Formats and File Formats*

5.5.2.1 Data Dissemination and Impediments

Presentation of virtual specimens in palaeontological publications has relied primarily on two-dimensional images or stereo pairs, in some cases supplemented by pre-rendered animations of the specimens being rotated and/or dissected (e.g. Sutton et al. 2001; Kamenz et al. 2008; Garwood and Sutton 2010). Static images can be high definition, but even where stereo pairs are used they cannot show an entire morphology. Pre-rendered animations can reveal the three-dimensional nature of morphology more clearly, but normally lack sufficient resolution to show all required detail. Thus, neither can be used as the basis for a full re-examination of a specimen; both provide a direct and convenient means of presenting specimens, but are a poor substitute for the three-dimensional data itself. The dissemination of three-dimensional morphological data underlying 'virtual palaeontology' publications is clearly desirable in the interests of scientific clarity, as well as to facilitate further research on the specimens (see e.g. Callaway 2011). Just as gene-sequence data underlying published (and unpublished) work are routinely made available to all interested parties via GenBank (www. ncbi.nlm.nih.gov/genbank), with immeasurable benefits for genetic science (Strasser 2008), the routine publication of virtual-fossil data would be beneficial for the science of palaeontology. Despite their desirability, however, such data releases have taken place only rarely in palaeontology (though see e.g. Rahman et al. 2012; Sutton et al. 2012b). Sutton et al. (2012a) considered that this partly reflects cultural impediments, that is, a reluctance amongst researchers to 'give away' data without any guarantee of reciprocation. Technical impediments, however, also exist.

Virtual palaeontological data can take many different forms depending on the approach used, and most workflows produce data in more than one form (Figure 5.1). Of the dataset types shown in Figure 5.1, perhaps five could reasonably be taken as construing a virtual specimen (surface point-clouds, registered tomographic datasets, prepared registered tomographic datasets, vector tomographic datasets and triangle meshes). Ideally, all publications making use of virtual palaeontological techniques would be accompanied by release of *all* datasets generated as part of the workflow, so that all processes and assumptions are made transparent and repeatable, but this is currently impractical for two reasons. Firstly, tomographic datasets in particular can be prohibitively large for dissemination, often exceeding 10 GB. While they can be reduced in size with lossy compression algorithms and/or downsampling (e.g. presented as videos, as by Siveter, Derek J., et al. 2004; Donoghue et al. 2006), this degrades them to the extent that they cannot be used as a basis for reconstructions. Permanent hosting of gigabyte-scale datasets is not technically problematic, but is relatively expensive; datasets of this size cannot routinely be hosted by journals as supplementary information, and while other online repositories for such data exist, their long-term stewardship may be problematic. Secondly, most dataset types that researchers might wish to release lack a widely acceptable file format that is not tied to a particular software package (e.g. prepared registered tomographic dataset), or can be represented by so many different file formats that no agreement on the 'correct' one has yet been reached (e.g. triangle-mesh datasets).

5.5.2.2   Triangle Meshes

As is apparent from Figure 5.1, there is a 'sweet spot' for data dissemination – three-dimensional triangle meshes. The vast majority of workflows discussed in this book either generate triangle meshes already or, in the case of point-cloud data, can easily be modified to do so (though see Section 5.3); the only exception is direct volume visualization (Section 5.4.3), which is rarely encountered. Triangle meshes are also relatively compact and homogeneous. Regardless of the preceding methodology, this data type comprises nothing more than a list of triangles defined by three points in space; some comprise multiple objects, but these are simply multiple lists of triangles with a limited quantity of metadata attached (e.g. object names). Triangle meshes are also 'visualization ready' – they require no specialized processing prior to display. For these reasons, we reiterate the recommendation of Sutton et al. (2012a) that virtual palaeontological specimens should normally be released as triangle-mesh datasets. Ideally, these should be supplemented by all datasets generated by earlier stages of the reconstruction methodology, but most scientists wishing to access virtual-fossil data will have no interest in revisiting the reconstruction process, but will simply want to carefully inspect the model that directly underlies published images, descriptions and interpretations. In most cases, triangle-mesh datasets provide precisely that model.

Table 5.3 Strengths and weaknesses of important triangle-mesh file formats.

Name	Advantages	Problems
STL & PLY	Simple file formats, widely used and understood by almost all software. Two subtypes – human readable (ASCII) format, computer readable (binary) format.	**No capacity for multiple objects; no capacity for storage of metadata;** ASCII STL files are very large; binary files smaller but lack compression facilities. STL files cannot store vertex colour information. PLY files can include non-standard attributes that not all software can read.
DXF	Simple file format, widely used and understood by most software. Human readable. Multiple named objects supported.	Files typically very large (larger than STL); **very limited facilities to store appropriate metadata** (e.g. no facility to correctly represent colour of objects).
3DS	Flexible format, compact, allows for some accompanying metadata.	**Limit of 65536 triangles per mesh; lacks facilities for arbitrary metadata tagging of objects;** not human readable.
VRML/X3D	Widely used format, though not as extensively so as STL and DXF; VRML (older iteration) human readable, some metadata facilities; X3D provides more compact binary files at expense of human readability.	VRML files in particular typically very large (larger than DXF); **viewing software relatively low performance (and X3D software scarce),** do not scale well to large triangle-count models; **lack facilities for arbitrary metadata tagging of objects.**
PDF/U3D	Three-dimensional PDFs (incorporating U3D) data can be viewed in free Adobe Reader software, already deployed to the bulk of computers; good metadata facilities; relatively small file sizes.	Limited range of viewing software (limited essentially to Adobe Reader) that lacks key facilities (e.g. stereo-viewing), **does not support export of data,** and performs poorly for complex models; limited free export tools to generate files; **lack of transparency and human readability in file format.**
VAXML (Sutton et al. 2012a)	Palaeontologically appropriate and simple metadata system that is human readable and writable; uses widely readable formats (STL/PLY) for triangle meshes themselves; extendable to point-cloud datasets through use of PLY files.	Datasets are multi-file (VAXML file plus one or more STL/PLY files), complicating delivery; not as compact as some formats; lack of direct viewing software (limited to SPIERSview) – although STL/PLY files can be imported into any software, metadata cannot at present.

Modified from Sutton et al. (2012a, Table 1). Reproduced with permission of The Palaeontological Association.
Weaknesses in bold face are those that we consider most problematic.

Many file formats exist that can store triangle-mesh datasets, but few are well-suited to the dissemination of virtual palaeontological specimens. Requirements (see Sutton et al. 2012a) include: simplicity (to ensure that all software using the format will understand all variants of files); transparency and 'openness' (data should be easy to generate and read, be human-readable as far as possible, and the format should not be tied to proprietary software); capacity to handle multiple objects; capacity to store appropriate metadata (object names, scales, colours, taxonomic names, authorship, etc.); compactness (file sizes should be as small as possible and single datasets should be represented by single files) and availability of import/export filters for reconstruction and visualization software. Strengths and weaknesses of some of the more commonly encountered formats not tied to proprietary software are listed in Table 5.3. Of these, the VAXML format proposed by two of us (Sutton et al. 2012a) is probably the best available solution to the file-format problem at present. VAXML datasets use a multi-file paradigm, where triangle data for models are provided in an existing and widely used format (STL or PLY), using multiple files if multiple objects are involved, supplemented with a single human-readable XML file providing metadata. Weaknesses of this format (see Table 5.3) are merely inconveniences; all other file formats that we are aware of have at least one issue that we consider to be substantially more serious (bold in Table 5.3). The SPIERS software website (www.spiers-software.org) provides examples of VAXML datasets for download and inspection.

5.6   Case Studies

We detail three different workflows for reconstruction and visualization of tomographic data; these have not been selected with a view to recommending any particular methodology, but are provided as exemplars of the details and considerations involved in reconstruction. The first of these is the methodology with which the authors have most familiarity, and is hence provided in the most detail.

5.6.1   The Herefordshire Lagerstätte (Isosurfacing; SPIERS; Physical-Optical)

The Silurian Herefordshire Lagerstätte preserves a diverse fossil fauna of soft-bodied invertebrates in three-dimensional form, and has been studied using physical-optical tomography; see Section 2.2.4.1 for more details of the deposit and of the data-capture methodology employed. Virtual reconstruction work on specimens from this deposit has been underway for over 11 years, and details of the workflow employed have changed. We describe here the current approach (e.g. of Sutton et al. 2012b) that uses the SPIERS software package (Section 5.5.1.1); see Sutton et al. (2001) for an earlier and less refined version of the methodology that pre-dates the development of SPIERS, making use instead of several proprietary software packages.

Herefordshire datasets from serial-grinding take the form of a series of evenly spaced tomographic photographs in '.bmp' format, numbered sequentially, and typically comprising 200–400 tomograms. Each specimen normally consists of two or more such datasets, as part and counterpart are ground separately, and large specimens are cut into sections for grinding (Section 2.2.4.1). The steps in reconstruction are as follows.

1   **Registration**. Datasets are registered (aligned) using the SPIERSalign program. Fiduciary edges, normally two, are used as

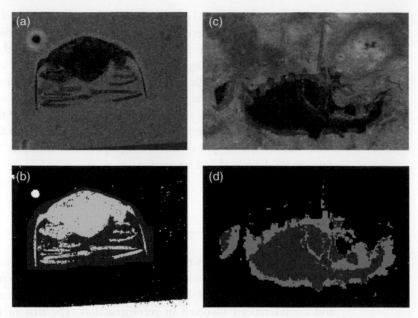

Figure 5.11 (a), (b) Tomogram of *Offacolus kingi* (OUMNH C.29558, figured in Sutton et al. 2002, Figure 2). Specimen is approximately 2 mm in width. (c), (d) Tomogram of *Bdellacoma* sp. (OUM C.29572, figured in Sutton et al. 2005, Fig. 2). Specimen is approximately 6 mm in width. (a) Cropped colour tomographic photograph. (b) Tomogram from SPIERSedit, reconstruction stage 5 (Section 5.6.1), downsampled by a factor of 2 (resolution is half that of a), monochrome, thresholded and with an initial mask defining a tight region of interest. (c) Cropped colour tomographic photograph. (d) Tomogram from SPIERSedit, reconstruction stage 6 (Section 5.6.1), downsampled by a factor of 2 (resolution is half that of a), monochrome, thresholded, and split into three segments: background (black), stereom (beige) and organic material (purple).

guides. Registration (see Section 5.3.2) is normally carried out manually, beginning in the middle of the dataset as tomograms at the ends are often partially obscured by resin, and proceeding first backwards then forwards, or vice versa. Some cleaner datasets are instead registered using automatic first-pass registration (Section 5.3.2). Tomogram comparison is achieved by rapid flicking between tomogram pairs; 'difficult' tomograms are temporarily hidden and registration carried out around them – they are later unhidden and slotted evenly into the sequence. Tomograms from near each end of the dataset that may lack fiduciary markings are registered by eye. Registration by a skilled operator proceeds at around 200 tomograms per day, and hence datasets can normally be registered in one or two days.

2   **Cropping**. After registration, datasets are cropped (see Section 5.2.1) to the smallest size possible, removing fiduciary edges where practical. Cropping is carried out using SPIERSalign, which outputs a cropped and registered colour tomographic dataset. This stage normally requires only a few minutes. Figure 5.11a, c shows cropped

tomograms after this stage of reconstruction. Resolution at this stage is typically 500–1000 pixels in each axis.

3 **Initial thresholding**. Datasets from stage 2 are imported into SPIERSedit. An initial downsampling level is chosen to ensure that the volume is close enough to isotropic (see Table 5.2) for isosurface reconstruction to be practical. In most cases, datasets are downsampled or 'binned' by a factor of 2; some such downsampled datasets are later upsampled again to generate detailed reconstructions of particular regions. Initial 'slice generation' is then performed. SPIERSedit does not use a variable threshold level, instead generating a monochrome 'working image' for each tomogram from a raw colour tomogram using 'slice-generation' rules; the user adjusts these rules such that the fixed threshold is optimal, on a per-tomogram basis if necessary (e.g. if brightness levels are different in different parts of the dataset). This stage normally requires only a few minutes.

4 **Initial visualization and assessment**. A visualization of the dataset is performed using SPIERSview. At this stage, an initial assessment of the specimen and its scientific potential is carried out; some specimens are abandoned at this point and not worked up any further. In some cases, the raw model is too noisy to visualize effectively – either the isosurface has too high a triangle count for rendering, or the level of noise obstructs viewing of the specimen, even with the use of island removal mesh processing (see Section 5.4.2.1). In these cases, initial assessment is deferred until after stage 5.

5 **Initial masking**. An initial masking (Section 5.2.4.3) pass is performed to separate the portions of the dataset that contain specimen from those that do not, enabling most noise and spurious data not directly adjacent to the specimen to be excluded from the reconstruction. Masking is carried out either by simply brushing the region of interest onto each tomogram using the mask-drawing tools of SPIERSedit, or more efficiently by using spline curves to manually define a region of interest at intervals throughout the dataset, and then interpolating between these. This initial masking pass normally takes no more than an hour. Figure 5.11b shows a tomogram from a specimen at this stage of reconstruction.

6 **Multiple segmentation.** Some specimens (e.g. Figure 5.11c) contain more than one type of fossil material, which can be automatically assigned to multiple segments (see Section 5.5.1.1) on the basis of colour in the original tomographic image. Where appropriate, multiple segments are identified at this stage using the segment generation tools of SPIERSedit (Figure 5.11d).

7 **Editing**. Specimens are subjected to editing (voxel modification; Section 5.3.4.3) on a tomogram-by-tomogram basis, using overlays of the original colour tomogram to ensure maximal accuracy. Figure 5.4 shows the difference between an unedited thresholded tomogram (Figure 2.4c) and an edited one (Figure 2.4e). This work

is performed in two dimensions, but regularly checked in three dimensions by rendering the part of the specimen being worked on in SPIERSview to ensure that no mistakes have been made, and that interpretation is consistent between tomograms. Editing work is manual, often painstaking, and can be slow; editing of each tomogram can take a skilled user anything from 2 to 30 minutes, depending on specimen complexity, the levels of noise present and the degree of 'fussiness' desired. Editing an entire dataset can thus represent anything from 1 to 25 days of work.

8 **Structural masking**. Once edited, specimens are masked into objects that will be reconstructed as discrete structures; Figure 5.5 shows an example. As for stage 5, the process can involve manual brushing of masks onto each tomogram, or for simpler regions the interpolation between spine curves. Masking is often an iterative process: one or two structures are identified and masked; a visualization is carried out; the masked structures are hidden; and further maskable structures identified. Once all maskable structures have been picked off in this way, only the remaining material will be left in the initial mask of stage 5; typically, this is the body of the organism. Often maskable structures (such as arthropod appendages) must be arbitrarily terminated where they meet another structure (such as the body). Care is taken that these arbitrary terminations are made consistently between tomograms (often with the aid of spline curves) to avoid 'ragged' structure endings. Note that stages 7 and 8 are not necessarily carried out discretely; many users prefer to undertake them in parallel.

9 **Production of research model**. The segmented, masked and edited model is 'polished' in SPIERSview by applying mesh processing (typically smoothing and a degree of mesh simplification, in some cases also island removal) and supplying some metadata (object names and grouping hierarchies, and a consistent colour scheme). Where the fossil comprises multiple datasets, these are brought together by importing all into one SPIERSview file, and manually moving each piece into the correct relative position. A final SPIERSview file is then distributed to the researchers, who use it to undertake morphological study of the specimen, typically using stereoscopic viewing to augment their interaction with the model. Herefordshire models in this form are typically between 3 and 15 million triangles in size.

10 **Export and ray tracing of model**. As preparation for publication, the model is exported from SPIERSview as a VAXML/STL dataset (see Sutton et al. 2012a). This 'canonical' version of the dataset can be used directly as an accompaniment to any publication. The STL files are also imported into Blender (Section 5.5.1.5). This package is used to produce ray-traced static images and stereo-pair images for published figures, as well as ray-traced animations of the specimen spinning in space, with or without partial dissections.

5.6.2   Caecilian Amphibians (Isosurfacing; Amira; Synchrotron CT)

Kleinteich et al. (2008) describe the production of three-dimensional visualizations and three-dimensional printed models of extant vertebrates from synchrotron CT data, using the commercial Amira package (www.vsg3d.com). While their methodology was applied to recent material rather than fossils, it provides a concise description of a typical workflow for CT data using commercial software. The data with which Kleinteich et al. (2008) were working were acquired at beamlines W2 and BW2 at the German Electron Sycnhrotron (DESY), and consisted of a stack of 8-bit TIFF-format images. These datasets were provided in variably downsampled (binned) form, by factors of 2, 3 and 4. The steps in reconstruction methodology are as follows:

1   **Virtual preparation**. Datasets were imported into Amira, where they were segmented and masked; Kleinteich et al. (2008) use the same term, segmentation, for both of these processes. They first performed what they termed segmentation (identification of material through different levels of grey in the image, i.e. different X-ray attenuation, as for SPIERS), and subsequently applied Amira 'labels' (= masks herein) to identify discrete structures; this was accomplished through tomogram-by-tomogram voxel-selection techniques, employing interpolation to speed up the process.

2   **Isosurfacing**. Labels and segmentation were used to produce separate volume datasets for different tissues. Isosurfaces for each of these were separately calculated in Amira, and exported in VRML format.

3   **Mesh processing**. VRML meshes were imported into the commercial ray tracing and animation package Autodesk Maya (usa.autodesk.com/maya); this involved the use of the conversion tool wrl2ma, as well as manual editing of VRML files to correct lighting and texture information. Maya was used to simplify the mesh to approximately 25% of its original triangle count; Kleinteich et al. (2008) do not record the algorithm used or the absolute triangle counts involved.

4   **Rendering and animation**. Ray-traced rendering (see Figure 5.12a) was performed in Maya, as was animation; these authors used the animation facilities in this package to analyze the relative movements of upper and lower jaw.

5   **Three-dimensional printing**. Models at 10–15 times life size were produced on a ZPrinter 310 rapid prototyping machine (ZCorp, Burlington, MA); see Figure 5.12b. Data were imported to the ZPrinter system in VRML format (see stage 2).

6   **Morphometric analysis**. Kleinteich et al. (2008) used the Amira 'surfacearea' tool, combined with measurements made from Maya models, to estimate muscle volume and hence strength. They also

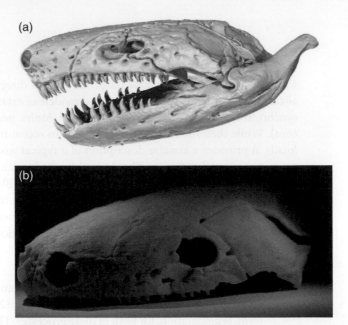

Figure 5.12 Caecilian skulls reconstructed using synchrotron CT data. (a) Ray-traced visualization of skull of *Siphonops annulatus*. (b) Three-dimensional printed model of *Ichthyophis* cf. *kohtaoensis* created using ZPrinting (Zprinter 310). Source: Kleinteich, T., Beckmann, F., Herzen, J., Summers, A. P. & Haas, A. 'Applying X-ray tomography in the field of vertebrate biology: form, function, and evolution of the skull of caecilians (Lissamphibia: Gymnophiona)', Developments in X-ray tomography VI. Proc. SPIE, 7078, 70780D, (2008). Reproduced with permission of SPIE and T. Kleinteich.

used oblique slices generated by Amira as the basis for measuring muscle fibre angles, using the two-dimensional image analysis capabilities of ImageJ (Section 5.5.1.3).

5.6.3 Neoproterozoic Problematica (Vector Surfacing; Scripting; Physical-Optical)

Maloof et al. (2010) describe an unusual vector-surfacing (Section 5.2.3) methodology to reconstruct weakly calcified fossils, around 5 mm in size, from Neoproterozoic bioclastic limestones. Their methodology is a refinement of that of Watters and Grotzinger (2001), to which readers are referred for a more detailed description of a subtly different vector-surfacing method.

The dataset of Maloof et al. (2010) was generated by serial grinding (at 50.8 μm intervals) and surface photography, and consisted of 470 tomograms. Fiduciary markings were present in the form of vertical holes drilled in each corner of their image. Their methodology did not involve a dedicated reconstruction package, but instead made use of a variety of commercial and free software packages. All work was automated through

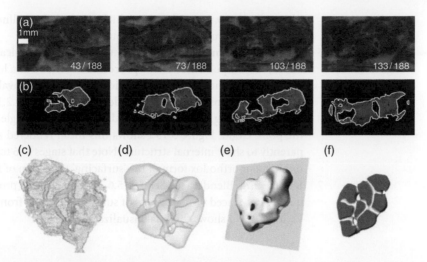

Figure 5.13 Reconstruction methodology of Maloof et al. (2010). (a) Four raw tomograms, white numbers indicate tomogram number in a sequence of 188 images. (b) Tomograms after contrast enhancement and auto-tracing. (c) Point-cloud representation of stacked auto-traces. (d) Meshed representation of point-cloud after low-pass filter, rendered semi-transparently. (e) Meshed specimen ray traced in Blender. (f) Meshed specimen sliced and ray traced using Blender. Source: Maloof et al. (2010, Fig. 3). Adapted by permission of Macmillan Publishers Ltd, copyright 2010.

scripts (written in Matlab and Visual Basic), such that they were able to generate reconstructions in no more than an hour per specimen. Their workflow was as follows:

1 **Registration.** Maloof et al. (2010) do not provide full documentation of their registration method other than to note that a Matlab script was used to 'cross-correlate and precisely orient each image'. Watters and Grotzinger (2001) used a manual registration process involving the measurement of the positions of fiduciary holes prior to tracing, and the use of these data to reposition traced contours later in the reconstruction process.

2 **Colour correction and cropping.** The commercial image-editing package Adobe Photoshop (www.adobe.com) was used to colour-correct tomograms, greatly increase their contrast and crop them to size. Figure 5.13a shows raw tomograms and Figure 5.13b contrast-enhanced tomograms.

3 **Auto-tracing.** Outlines of fossils after contrast enhancement were converted to vector format using the 'autotrace' tool of the commercial vector-graphics package Adobe Illustrator (www.adobe.com), which uses edge-detection algorithms to find traceable lines in raster images. Figure 5.13b shows these traced splines overlaid on the tomograms. These vector objects were exported in DXF format.

4 **Conversion to point cloud.** Vector tracings from Adobe Illustrator were imported into the commercial three-dimensional modeller

Rhino3D (www.rhino3d.com), which treated the spline nodes as points, and the model as a point cloud (Figure 5.13c).

5 **Low-pass filtering**. Rhino3D was used to algorithmically remove outliers of the point cloud; this process is described as a low-pass filter, and performed the same functions as island removal and mesh smoothing in isosurface-based models (see Section 5.4.2.1).

6 **Surface meshing**. Rhino3D was used to generate a triangle mesh from the point cloud; Figure 5.13d shows this mesh rendered semi-transparently to show internal structure. Note that stages 4–6 together represent an unorthodox form of the 'surfacing' transition of Figure 5.1.

7 **Ray tracing**. Blender (see Section 5.6.1.5) was used to produce high-quality ray-traced visualizations of surfaces exported from Rhino3D. Figure 5.13e, f shows the final visualization.

References

Abel, R.L., Laurini, C.R. & Richter, M. (2012) A palaeobiologist's guide to 'virtual' micro-CT preparation. *Palaeontologia Electronica*, **15** (2), 6T.

Albani, A.E., Bengtson, S., Canfield, D.E., et al. (2010) Large colonial organisms with coordinated growth in oxygenated environments 2.1 Gyr ago. *Nature*, **466**, 100–104.

Bade, R., Haase, J. & Preim, B. (2006) Comparison of fundamental mesh smoothing algorithms for medical surface models. *Simulation und Visualisierung 2006 (SimVis 2006)*, pp. 289–304. SCS Verlag, Magdeburg.

Callaway, E. (2011) Fossil data enter the web period. *Nature*, **472**, 150.

Chapman, R.E. (1989) Computer assembly of serial sections. In: Feldmann, M., Chapman, R. & Hannibal, J.T. (eds), *Paleotechniques*, pp. 157–164. Special Publication 4, Paleontological Society, Boulder.

Donoghue, P.C.J., Bengtson, S., Dong, X., et al. (2006) Synchrotron X-ray tomographic microscopy of fossil embryos. *Nature*, **442**, 680–683.

Falkingham, P.L. (2012) Acquisition of high resolution 3D models using free, open-source, photogrammetric software. *Palaeontologia Electronica*, **15** (1), 1T.

Gai, Z., Donoghue, P.C.J., Zhu, M., et al. (2012) Fossil jawless fish from China foreshadows early jawed vertebrate anatomy. *Nature*, **476**, 324–327.

Garland, M. & Heckbert, P.S. (1997) Surface simplification using quadric Error metrics. *SIGGRAPH '97 Proceedings of the 24th Annual Conference on Computer Graphics and Interactive Techniques*, pp. 209–216. ACM, New York.

Garland, M. & Heckbert, P.S. (1998) Simplifying surfaces with color and texture using quadric error metrics. *Proceedings of Visualization 98*, pp. 263–269. IEEE Computer Society, New York.

Garwood, R., Dunlop, J., & Sutton, M. (2009) High-fidelity X-ray micro-tomography reconstruction of siderite-hosted Carboniferous arachnids. *Biology Letters*, **5**, 841–844.

Garwood, R.J. & Sutton, M.D. (2010) X-ray micro-tomography of Carboniferous stem-Dictyoptera: new insights into early insects. *Biology Letters*, **6**, 699–702.

Gibson, I., Rosen, D.W. & Stucker, B. (2010) *Additive Manufacturing Technologies: Rapid Prototyping to Direct Digital Manufacturing*. Springer, New York.

Glassner, A. (1989) *An Introduction to Ray Tracing*. Morgan Kauffman, San Francisco.

Herbert, M.J. (1999) Computer-based serial section reconstruction. In: Harper, D.A.T. (ed), *Numerical Palaeobiology: Computer-Based Modelling and Analysis of Fossils and Their Distributions*, pp. 93–126. Wiley, Chichester.

Kamenz, C., Dunlop, J.A., Scholtz, G., et al. (2008) Microanatomy of early Devonian book lungs. *Biology Letters*, **4 (2)**, 212–215.

Kleinteich, T., Beckmann, F., Herzen, J., et al. (2008). Applying X-ray tomography in the field of vertebrate biology: form, function, and evolution of the skull of caecilians (Lissamphibia: Gymnophiona). *Developments in X-ray tomography VI. Proceedings of SPIE*, **7078**, 70780D.

Lichtenbelt, B., Crane, R. & Naqvi, S. (1998) *Introduction to Volume Rendering*. Prentice Hall, Upper Saddle River.

Long, J.A., Trinajstic, K., Young, G.C., et al. (2008) Live birth in the Devonian period. *Nature*, **453**, 650–622.

Lorensen, W.E. & Cline, H.E. (1987). Marching cubes: a high resolution 3D surface construction algorithm. *Computer Graphics*, **(21) 4**, 163–169.

Maloof, A.C., Rose, C.V., Beach, R., et al. (2010) Possible animal-body fossils in pre-Marinoan limestones from South Australia. *Nature Geoscience*, **3**, 653–659.

Mallison, H. (2011). Digitizing methods for paleontology: applications, benefits and limitations. In: Elewa, A. (ed), *Computational Paleontology*. Springer-Verlag, Berlin.

Marton, Z.S., Rusu, R.B. & Beetz, M. (2009). On fast surface reconstruction methods for large and noisy point clouds. *2009 IEEE International Conference on Robotics and Automation, Kobe, Japan*, pp. 3218–3223. IEEE, New York.

Penny, D., Dierick, M., Cnudde, V., et al. (2007). First fossil Micropholcommatidae (Araneae), imaged in Eocene Paris amber using X-ray computed tomography. *Zootaxa*, **1623**, 47–53.

Rahman, I.A. & Clausen, S. (2009) Re-evaluating the palaeobiology and affinities of the Ctenocystoidea (Echinodermata). *Journal of Systematic Palaeontology*, 7 **(4)**, 413–426.

Rahman, I.A., Adcock, K. & Garwood, R.J. (2012) Virtual fossils: a new resource for science communication in paleontology. *Evolution: Education and Outreach*, **5**, 635–641.

Salman, N., Yvinec, M. & Merigot, Q. (2010) Feature preserving mesh generation from 3D point clouds. *Computer Graphics Forum*, **29 (5)**, 1623–1632.

Schiffbauer, J.D. & Xiao, S. (2009) Novel application of focused ion beam electron microscopy (FIB-EM) in preparation and analysis of microfossil ultrastructures: a new view of complexity in early eukaryotic organisms. *Palaios*, **24 (9)**, 616–626.

Schroeder, W.J., Zarge, J.A. & Lorensen, W.E. (1992) Decimation of triangle meshes. *SIGGRAPH '92 Proceedings of the 19th Annual Conference on Computer Graphics and Interactive Techniques*, pp. 65–70. ACM Press, New York.

Selden, P.A., Shear, W.A. & Sutton, M.D. (2008) Fossil evidence for the origin of spider spinnerets, and a proposed arachnid order. *Proceedings of the National Academy of Sciences*, **105 (52)**, 20781–20785.

Siveter, Derek J., Briggs, D.E.G., Siveter, David J., et al. (2004) A Silurian sea spider. *Nature*, **431**, 978–980.

Stegmaier, S., Strengert, M., Klein, T., et al. (2005) A simple and flexible volume rendering framework for graphics-hardware-based raycasting. *VG'05 Proceedings of the Fourth Eurographics/IEEE VGTC conference on Volume Graphics*, pp. 187–195. Eurographics Association, Aire-la-Ville.

Strasser, B.J. (2008) GenBank – Natural history in the 21st century? *Science*, **322**, 537–538.

Sutton, M.D., Briggs, D.E.G., Siveter, David J., et al. (2001) Methodologies for the visualization and reconstruction of three-dimensional fossils from the Silurian Herefordshire Lagerstätte. *Palaeontologia Electronica*, **4 (1)**, 1A.

Sutton, M.D., Briggs, D.E.G., Siveter, David J., et al. (2002) The arthropod Offacolus kingi (Chelicerata) from the Silurian of Herefordshire, England: computer based morphological reconstructions and phylogenetic affinities. *Proceedings of the Royal Society B*, **269**, 1195–1203.

Sutton, M.D., Garwood, R.J., Siveter, David J., et al. (2012a) SPIERS and VAXML; a software toolkit for tomographic visualisation and a format for virtual specimen interchange. *Palaeontologia Electronica*, **15 (2)**, 5T.

Sutton, M.D., Briggs, D.E.G., Siveter, David J., et al. (2012b) A Silurian armoured aplacophoran and implications for molluscan phylogeny. *Nature*, **490 (7418)**, 94–97.

Watters, W.A. & Grotzinger, J.P. (2001) Digital reconstruction of calcified early metazoans, terminal Proterozoic Nama Group, Namibia. *Paleobiology*, **27**, 159–171.

Wimmer, M. & Scheiblauer, C. (2006) Instant points: fast rendering of unprocessed point clouds. *SPBG'06 Proceedings of the 3rd Eurographics/IEEE VGTC conference on Point-Based Graphics*, pp. 129–137. Eurographics Association, Aire-la-Ville.

Whybrow, P.J. & Lindsay, W. (1990) Preparation of macrofossils. In: Briggs, D.E.G. & Crowther, P.R. (eds), *Palaeobiology: A Synthesis*, pp. 500–502. Blackwell, Oxford.

Young, M.T., Rayfield, E.J., Holliday, C.M., et al. (2012) Cranial biomechanics of *Diplodocus* (Dinosauria, Sauropoda): testing hypotheses of feeding behaviour in an extinct megaherbivore. *Naturwissenschaften*, **99**, 637–643.

Zamora, S., Rahman, I.A. & Smith, A.B. (2012) Plated Cambrian bilaterians reveal the earliest stages of echinoderm evolution. *PLos ONE*, **7 (6)**, e38296.

Further Reading/Resources

Hansen, C.D. & Johnson, C.R. (2011) *Visualization Handbook*. Academic Press, Salt Lake City.

Hansen, G.A., Douglas, R.W. & Zardecki, A. (2005) *Mesh Enhancement: Selected Elliptic Methods, Foundations and Applications*. Imperial College Press, London.

Hess, R. (2010) *The Essential Guide to Learning Blender 2.6*. Elsevier, Amsterdam.

Preim, B. & Bartz, D. (2007) *Visualization in Medicine: Theory, Algorithms, and Applications*. Morgan Kauffman, Burlington.

6

Applications beyond Visualization

Abstract: Virtual fossils have great value as a means of quantifying form and function, and can be used to address a number of different questions in palae-ontology. Geometric morphometrics uses topologically homologous land-mark points on digital reconstructions to characterize shape variation among fossils (e.g. vertebrate skulls). Dental microwear texture analysis measures microscopic wear patterns on digitized tooth surfaces; this can help recon-struct dietary preferences, for example in fossil mammals. Several computer modelling methods have proven useful for palaeobiomechanics. Finite-element analysis reconstructs stress and strain in virtual reconstructions of skulls during biting. Multibody dynamics analysis is a method for simulating jaw movements, which can be applied to extinct species. Body-size estimation utilizes computer models to estimate the dimensions of vertebrates, including dinosaurs. Finally, computational fluid dynamics simulates fluid flow (air or water) past fossils, informing on their aero- or hydrodynamics.

6.1 Introduction

In previous chapters, we have detailed various techniques for digitizing fossils, including approaches for both acquiring (Chapters 2–4) and visual-izing (Chapter 5) three-dimensional datasets. The case studies presented in these chapters demonstrate the utility of virtual fossils in palaeontology, especially for anatomical descriptions. Such computer reconstructions can also serve as a basis for quantitative analyses of form and function. Here, we provide a brief overview of three applications beyond visualization: geometric morphometrics (Section 6.2), **dental microwear texture analysis** (Section 6.3) and biomechanical modelling (Section 6.4). This is not intended to be an exhaustive list of all potential applications, but rather introduces some of those most commonly applied in palaeontology.

Techniques for Virtual Palaeontology, First Edition. Mark D. Sutton, Imran A. Rahman and Russell J. Garwood.
© 2014 John Wiley & Sons, Ltd. Published 2014 by John Wiley & Sons, Ltd.

6.2   Geometric Morphometrics

Morphometrics is the quantitative study of form. Traditional morphometrics involves the analysis of simple measurements (e.g. distances, ratios and angles) of morphological structures. Such values are typically strongly correlated with size, however, and do not encode information about the location of the measurements. **Geometric morphometrics** was developed in the 1980s as a solution to these problems; it uses data that capture object geometry (Rohlf and Marcus 1993; Adams et al. 2004). Such data consist of **landmarks**, which are topologically homologous points in all studied individuals. Landmark configurations have traditionally been analyzed in two dimensions; however, they are increasingly being analyzed in three dimensions using digital reconstructions produced with tomographic or surface-based methods. Alternatively, mechanical digitization enables direct collection of three-dimensional landmark data, simplifying the object to a configuration of points on a computer (Section 4.4). In addition to true anatomical landmarks, outlines and surfaces can be represented by arbitrary points constrained onto them, which are called semi-landmarks. The effect of size can be removed through a process called Procrustes superimposition (Rohlf 1999); here, objects are superimposed by translation, rotation and scaling (sliding for semi-landmarks), making points comparable between specimens. Subsequently, comparative plots and statistical analyses enable quantitative exploration of shape variability among individuals in a morphometric dataset.

In palaeontology, three-dimensional morphometrics has been applied predominantly to fossil vertebrates, especially skulls (e.g. Gunz et al. 2009; Goswami et al. 2011; O'Higgins et al. 2011). Souter et al. (2010) used landmark-based geometric morphometrics to study three-dimensional shape variation in avian and extinct theropod pelves. This study created computer reconstructions of four modern birds through X-ray microtomography and a reconstruction of the fossil *Allosaurus fragilis* through laser scanning. Utilizing a reference template – a simple model of a theropod pelvis built in the graphics package Maya (usa.autodesk.com/maya/) – they were able to place 44 true landmarks and 892 semi-landmarks (curves and surfaces) on these virtual pelves (Figure 6.1). This was achieved

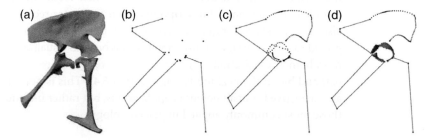

Figure 6.1 Geometric morphometrics of the pelvis of the theropod dinosaur *Allosaurus fragilis*. (a) Digital reconstruction. (b) Configuration of true landmarks. (c) Configuration of true landmarks with curve semi-landmarks. (d) Configuration of true landmarks with surface semi-landmarks. Source: Souter et al. (2010, Fig. 1). Reproduced with permission of Elsevier.

using the program Edgewarp3D (vhp.med.umich.edu/edgewarpss.html), which allows warping of a reference configuration of landmarks and semi-landmarks onto a three-dimensional digital object. Data were treated with Procrustes superimposition to remove scaling effects, and a principal component analysis was carried out to reveal shape variation among the theropod taxa.

6.3 Dental Microwear Texture Analysis

Dental microwear analysis is an approach for inferring the diets of animals based on microscopic wear patterns on teeth. Conventionally, this was performed by examining two-dimensional occlusal surfaces (i.e. those used for biting or chewing) with scanning electron microscopy (e.g. Walker et al. 1978; Teaford and Walker 1984); however, the last decade has seen the development of a new technique for studying microwear in three dimensions: dental microwear texture analysis (Ungar et al. 2003; Scott et al. 2006). Here, teeth – fossil or recent – are imaged three-dimensionally at sub-micrometre resolutions using, for example, confocal microscopy. Surface texture variables (e.g. directionality and roughness) are then automatically and quantitatively characterized with scale-sensitive fractal analysis (SSFA) – a method for measuring small-scale features on three-dimensional surfaces. This approach enables the documentation of characters that cannot easily be observed or quantified in two dimensions, and is more objective and precise than traditional techniques, in which surface features (i.e. pits and scratches) are manually scored by an operator.

Palaeontological studies employing dental microwear texture analysis have focused almost exclusively on fossil mammals (e.g. Prideaux et al. 2009; Ungar et al. 2010). Scott et al. (2005) applied the technique to a variety of living primates and extinct hominins (including *Australopithecus africanus* and *Paranthropus robustus*). They used a Sensofar PLμ confocal microscope to scan four adjacent fields of view (138 × 102 μm) on each studied tooth, giving a total sampled area of 276 × 204 μm. The resulting three-dimensional surfaces (e.g. Figure 6.2) were then subjected to SSFA. Using the program Kfrax (www.surfract.com/), the area–scale fractal complexity (derived from a series of relative area measurements at different scales) and the exact proportion length–scale anisotropy of relief (derived from a series of relative length measurements at different orientations for a given scale) were measured. These values served to quantify microwear complexity and directionality (anisotropy), respectively. Finally, statistical analyses were performed to test for differences between taxa, revealing that microwear was more complex in *P. robustus* and more anisotropic in *A. africanus*; this suggests that *A. africanus* ate more tough foods than *P. robustus*.

102 μm   139 μm

Figure 6.2 Three-dimensional occlusal surface of a tooth of the hominin *Australopithecus africanus* used for dental microwear texture analysis. Colours indicate surface topography: blues and greens represent regions of relatively low elevation; reds and yellows represent regions of relatively high elevation. Source: Scott et al. (2005, Fig. 2). Reproduced with permission of Nature Publishing Group.

6.4 Biomechanical Modelling

A number of different modelling approaches – co-opted from engineering – have been used to quantitatively explore the link between form and function in three-dimensional virtual fossils (Figure 6.3, Figure 6.4, Figure 6.5 and Figure 6.6). Models are simplifications that facilitate investigation of real, complex biological systems; wherever possible, sensitivity analyses should be performed to test the influence of different input parameters (e.g. soft-tissue reconstructions), and models validated using, for example, experimental data from appropriate extant taxa (Anderson et al. 2012; Hutchinson 2012). Four computational approaches are summarized below: finite-element analysis, **multibody dynamics analysis, body-size estimation** and computational fluid dynamics. These methods can be applied independently or, in some cases, combined for integrative analyses of functional morphology. The growing availability and power of high-performance computers is enabling increasingly rapid and detailed analyses.

6.4.1 Finite-Element Analysis

Finite-element analysis (FEA) is a technique for reconstructing stresses and strains in geometrical objects. By integrating realistic physical properties into models, structures such as bone can be studied. Models can be one-, two-, or three-dimensional, depending on the questions being asked. Originally applied in mechanical engineering (Zienkiewicz 1971), FEA was subsequently used in orthopaedics (Huiskes and Chao 1983) and, later, in biomechanical studies of extant and extinct animals (e.g. Carter et al. 1998; Rayfield et al. 2001). For three-dimensional applications, specimens are digitized with appropriate methods (e.g. medical CT for large vertebrate skulls) and transformed into models consisting of large yet finite numbers of discrete sub-regions or elements. Material properties (e.g. of bone) and **boundary conditions** (i.e. constraints and loads) are applied to the

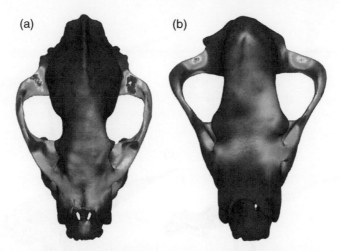

Figure 6.3 Finite-element analysis (FEA) models of extant and fossil hyenas. (a) The skull of the extant spotted hyena *Crocuta crocuta*. (b) The skull of the fossil hyena-like carnivore *Dinocrocuta gigantea*. Colours indicate stress distribution during a bite: blues and greens represent regions of relatively low stress; reds and yellows represent regions of relatively high stress. Source: Tseng (2009, Fig. 2). Reproduced with permission of John Wiley and Sons.

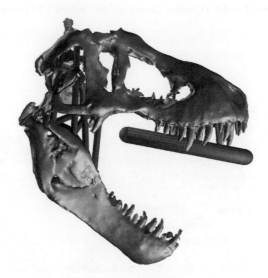

Figure 6.4 Multibody dynamics analysis (MDA) model of the skull of the theropod dinosaur *Tyrannosaurus rex* with simplified muscles (red) and contact springs (blue) to measure bite force. Source: Bates and Falkingham (2012, Fig. 1). Reproduced with permission of Royal Society Publishing.

models – in the case of extinct species, these input parameters might be inferred from closely related modern taxa (which are identified using extant phylogenetic bracketing; Witmer 1995). The results of FEA can be represented visually as scaled colour contour plots of deformation/stress/strain magnitudes; alternatively, quantitative data can be extracted from individual

Figure 6.5 Body-size estimation modelling of the theropod dinosaur *Tyrannosaurus rex*. (a) Digital reconstruction of the skeleton. (b) Digital reconstruction with elliptical sections to define body outline. (c) Digital reconstruction with air spaces added. (d) Final meshed reconstruction. *Source*: Hutchinson et al. (2011, Fig. 1). Reproduced from the original image, which is licensed under a Creative Commons Attribution 2.5 License (http://creativecommons. org/licenses/by/2.5/).

Figure 6.6 Computational fluid dynamics (CFD) model of the trilobite *Hypodicranotus striatus* showing flow velocity vectors for different inlet velocities (0.01, 0.2 and 0.5 m/s) and morphologies. (a) Model with hypostome. (b) Model without hypostome. Source: Shiino et al. (2012, Fig. 4). Reproduced with permission of Elsevier.

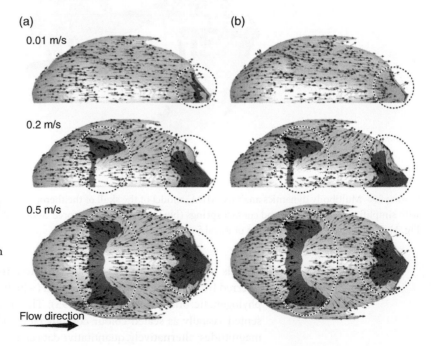

elements or nodes, providing absolute values of deformation, stress and strain within the structure during loading.

In recent years, three-dimensional FEA has been utilized in studies of a range of fossil vertebrates (e.g. Degrange et al. 2010; Dumont et al. 2011; Jones et al. 2012). Where input parameters are poorly known, as is frequently the case for fossils, absolute values cannot be accurately estimated using FEA (Bright and Rayfield 2011). In such cases, the technique is better employed for comparative analyses of different bite scenarios and skull morphologies. Tseng (2009) applied comparative FEA to the skulls of a fossil and an extant hyena and an extant wolf. This study reconstructed the specimens using medical CT, and the FEA software STRAND7 (www.strand7.com) was used to create three-dimensional finite-element models of the digital skulls. Sensitivity analyses were conducted to determine the influence of varying input parameters, and the same material properties (Young's modulus = 20 GPa and Poisson's ratio = 0.3) – which were taken from published estimates for birds and mammals – were assigned to all three models. Models were constrained at three locations (two jaw joints and a tooth), and muscular loads were applied. Jaw muscles were modelled based on osteological correlates and muscle arrangements in extant relatives, and bite force was derived from experimental data for an extant hyena. The results demonstrate that the stress experienced during biting varied considerably with skull morphology (Figure 6.3).

6.4.2   Multibody Dynamics Analysis

Multibody dynamics analysis (MDA) is a method for modelling the movements of multiple interconnected bodies. In biomechanics, this approach is usually applied to the dynamics of the head and jaws, and to locomotion; complex three-dimensional models are becoming increasingly common in studies of craniofacial function (e.g. Langenbach et al. 2002; Sellers and Crompton 2004). In such analyses, the skull and lower jaw are digitally reconstructed in three dimensions, joints and soft tissues (i.e. muscles and ligaments) virtually modelled, and jaw movements simulated in specialist software. This allows the researcher to address questions relating to bone movements, muscle forces, joint kinematics and bite forces in vertebrates (modern or fossil). Moreover, the results of MDA (e.g. muscular loads and reaction forces) can be used as boundary conditions in FEA in order to reduce the need for constraints (see e.g. Moazen et al. 2008; Curtis et al. 2011).

MDA is highly reliant on accurate soft-tissue reconstructions, which necessitate numerous assumptions when modelling extinct species. To date, it has been employed rarely in palaeontology. Bates and Falkingham (2012) investigated the bite performance of the theropod dinosaur *Tyrannosaurus rex* with MDA. They used laser scanning to digitize the skull of an adult *T. rex* and CT to image the skulls of an adult human and a juvenile alligator. The jaw joint and main muscles were modelled, for all of these reconstructions, in the graphics package Maya (usa.autodesk.com/maya). Virtual fossils were then imported into the MDA package GaitSym (www.animalsimulation. org/page3/page7/page7.html); parameters (fibre length = 25% maximum muscle length; physiological cross-sectional area = best estimate of muscle volume divided by fibre length; maximum contraction velocity = 8 s^{-1}; and force per unit area = 300,000 N m^2) were assigned to the muscles of the *T. rex* model based on comparisons with the human and alligator models (for which muscle properties were derived from the literature). Modelling was validated by comparing the bite forces estimated by the human and alligator models to experimental data, and *T. rex* muscle parameters were varied as part of a **sensitivity analysis** to determine their influence on functional performance. Biting was simulated in *T. rex* with the jaws opened to 45° and muscles activated for one second (Figure 6.4), giving the highest bite force (35,000–57,000 N) estimated for any terrestrial animal.

6.4.3 Body-Size Estimation

Computer models are also used to more accurately estimate the dimensions of extinct vertebrates, especially dinosaurs, in order to better understand their life habits. Of particular importance are estimates of body size, as this aspect of organismal biology is critical to function. Early work involved the production of three-dimensional digital representations based on two-dimensional outline drawings of life reconstructions (Henderson 1999). More recently, imaging of complete, articulated skeletons with surface-based methods (e.g. laser scanning) has become the standard for three-dimensional visualization prior to modelling (e.g. Gunga et al. 2007; Bates et al. 2009; Hutchinson et al. 2011; Sellers et al. 2012). Whole-body outlines of the animal's soft tissues are reconstructed around the virtual skeleton in graphics software, with mass properties then calculated using the volume of the three-dimensional model and density estimates from the literature.

T. rex has been a major focus for model-based research on dinosaur body size (e.g. Henderson 1999; Hutchinson et al. 2007; Bates et al. 2009). Hutchinson et al. (2011) applied the three-dimensional procedure outlined above to *T. rex* body dimensions. This study imaged several well-preserved, nearly complete skeletons using laser scanning, mechanical digitization and CT. The resulting three-dimensional reconstructions were imported into computer graphics software, and elliptical sections were placed around each skeleton and surfaces lofted between sections to create the body outlines. Air spaces (representing the lungs and air sacs) were added as zero-density shapes in the head, neck and torso (Figure 6.5). Density was set at 1000 kg/ m^3 for each body segment and software was used to estimate the mass for segments and the whole body of each specimen. Sensitivity analyses were performed by varying the body outlines and assessing the impact this had on mass estimates. Modelling results suggested that the body mass of adult *T. rex* typically varied from about 6000 to 8000 kg.

6.4.4 Computational Fluid Dynamics

Computational fluid dynamics (CFD) entails the use of computer models to study the flow of fluids (both liquids and gases) around objects. Since the earliest two-dimensional numerical simulations in the 1930s, the discipline has primarily been concerned with addressing complex engineering-design problems, often in three dimensions (Shang 2004). However, CFD can also be used to analyze the functional performance of living and fossil organisms in aquatic and aerial settings (Kato and Kamimura 2008; Miller et al. 2012). A three-dimensional representation of the organism – based on a computer reconstruction (produced using, e.g., CT) or digitally modelled – is introduced into a virtual fluid-filled volume. Flow and boundary conditions (i.e. fluid behaviour and properties) are defined and the simulation is then performed, with the governing fluid dynamics equations solved numerically. During post-processing, the results are visualized and analyzed to quantify the influence of form on flow. Because object geometries and model input parameters can be easily varied, and modelling results simply visualized and quantified, CFD is much more flexible and powerful than traditional physical modelling. However, validation of the computer model using experimental data from a physical model in a flume tank or wind tunnel is important for testing the accuracy of computational modelling.

A handful of published studies have used CFD to examine function in fossil taxa (e.g. Rigby and Tabor 2006; Shiino et al. 2009; Shiino and Kuwazuru 2010). Shiino et al. (2012) employed the method when investigating the function of the hypostome (a long, fork-like structure near the mouth) in the trilobite *Hypodicranotus striatus*. They created polycarbonate replicas (with and without a hypostome) based on a complete, articulated specimen; these replicas were imaged with X-ray microtomography to obtain three-dimensional digital reconstructions. The reconstructions were imported into the CFD code SCRYU/Tetra (www.cradle.co.jp/products/scryutetra/index.html), where finite-element meshes of the virtual fossils were generated. Simulations were then performed in cylindrical virtual flume tanks with the models fixed in space. The Reynolds-averaged Navier–Stokes equations and $k - \varepsilon$ turbulence model were implemented, and the simulations were run for 15 seconds (time step of 0.003 seconds). In a preliminary analysis, mesh complexity was varied to establish how this influenced flow behaviour; optimal meshing parameters were determined in this way and used in subsequent modelling. Final simulations were undertaken using 11 different inlet velocities (0.01–0.5 m/s) and corresponding Reynolds numbers (270–13,500). The results were visualized in three dimensions (e.g. as flow velocity vectors; Figure 6.6) and analyzed to obtain values for the coefficients of drag and lift. This demonstrated that the presence of a hypostome in models served to reduce drag (presumably aiding swimming in the living animal).

References

Adams, D.C., Rohlf, F.J. & Slice, D.E. (2004) Geometric morphometrics: ten years of progress following the 'revolution'. *Italian Journal of Zoology*, **71 (1)**, 5–16.

Anderson, P.S.L., Bright, J.A., Gill, P.G., et al. (2012) Models in palaeontological functional analysis. *Biology Letters*, **8 (1)**, 119–122.

Bates, K.T. & Falkingham, P.L. (2012) Estimating maximum bite performance in *Tyrannosaurus rex* using multi-body dynamics. *Biology Letters*, **8 (4)**, 660–664.

Bates, K.T., Manning, P.L., Hodgetts, D., et al. (2009) Estimating mass properties of dinosaurs using laser imaging and 3D computer modelling. *PLoS ONE*, **4 (2)**, e4532.

Bright, J.A. & Rayfield, E.J. (2011) Sensitivity and *ex vivo* validation of finite element models of the domestic pig cranium. *Journal of Anatomy*, **219 (4)**, 456–471.

Carter, D.R., Mikić, B. & Padian, K. (1998) Epigenetic mechanical factors in the evolution of long bone epiphyses. *Zoological Journal of the Linnean Society*, **123 (2)**, 163–178.

Curtis, N., Jones, M.E., Shi, J., et al. (2011) Functional relationship between skull form and feeding mechanisms in Sphenodon, and implications for diapsid skull development. *PLoS ONE*, **6 (12)**, e29804.

Degrange, F.J., Tambussi, C.P., Moreno, K., et al. (2010) Mechanical analysis of feeding behavior in the extinct "terror bird" *Andalgalornis steulleti* (Gruiformes: Phorusrhacidae). *PLoS ONE*, **5 (8)**, e11856.

Dumont, E.R., Ryan, T.M. & Godfrey, L.R. (2011) The *Hadropithecus* conundrum reconsidered, with implications for interpreting diet in fossil hominins. *Proceedings of the Royal Society B*, **278 (1725)**, 3654–3661.

Goswami, A., Milne, N. & Wroe, S. (2011) Biting through constraints: cranial morphology, disparity and convergence across living and fossil carnivorous mammals. *Proceedings of the Royal Society B*, **278 (1713)**, 1831–1839.

Gunga, H.-C., Suthau, T., Bellmann, A., et al. (2007) Body mass estimations for *Plateosaurus engelhardti* using laser scanning and 3D reconstruction methods. *Naturwissenschaften*, **94 (8)**, 623–630.

Gunz, P., Mitteroecker, P., Neubauer, S., et al. (2009) Principles for the virtual reconstruction of hominin crania. *Journal of Human Evolution*, **57 (1)**, 48–62.

Henderson, D.M. (1999) Estimating the masses and centers of mass of extinct animals by 3-D mathematical slicing. *Paleobiology*, **25 (1)**, 88–106.

Huiskes, R. & Chao, E.Y. (1983) A survey of finite element analysis in orthopedic biomechanics: the first decade. *Journal of Biomechanics*, **16 (6)**, 385–409.

Hutchinson, J.R. (2012) On the inference of function from structure using biomechanical modelling and simulation of extinct organisms. *Biology Letters*, **8 (1)**, 115–118.

Hutchinson, J.R., Thow-Hing, V.N. & Anderson, F.C. (2007) A 3D interactive method for estimating body segmental parameters in animals: application to the turning and running performance of *Tyrannosaurus rex*. *Journal of Theoretical Biology*, **246 (4)**, 660–680.

Hutchinson, J.R., Bates, K.T., Molnar, J., et al. (2011) A computational analysis of limb and body dimensions in *Tyrannosaurus rex* with implications for locomotion, ontogeny, and growth. *PLoS ONE*, **6 (10)**, e26037.

Jones, D., Evans, A.R., Siu, K.K.W., et al. (2012) The sharpest tool in the box? Quantitative analysis of conodont element functional morphology. *Proceedings of the Royal Society B*, **279 (1739)**, 2849–2854.

Kato, N. & Kamimura, S. (eds) (2008) *Bio-Mechanics of Swimming and Flying*. Springer, Tokyo.

Langenbach, G.E.J., Zhang, F., Herring, S.W., et al. (2002) Modelling the masticatory biomechanics of a pig. *Journal of Anatomy*, **201 (5)**, 383–393.

Miller, L.A., Goldman, D.I., Hedrick, T.L., et al. (2012) Using computational and mechanical models to study animal locomotion. *Integrative and Comparative Biology*, **52 (5)**, 553–575.

Moazen, M., Curtis, N., Evans, S.E., et al. (2008) Combined finite element and multibody dynamics analysis of biting in a *Uromastyx hardwickii* lizard skull. *Journal of Anatomy*, **213 (5)**, 499–508.

O'Higgins, P., Cobb, S.N., Fitton, L.C., et al. (2011) Combining geometric morphometrics and functional simulation: an emerging toolkit for virtual functional analyses. *Journal of Anatomy*, **218 (1)**, 3–15.

Prideaux, G.J., Ayliffe, L.K., DeSantis, L.R.G., et al. (2009) Extinction implications of a chenopod browse diet for a giant Pleistocene kangaroo. *Proceedings of the National Academy of Sciences of the United States of America*, **106 (28)**, 11646–11650.

Rayfield, E.J., Norman, D.B., Horner, C.C., et al. (2001) Cranial design and function in a large theropod dinosaur. *Nature*, **409 (6823)**, 1033–1037.

Rigby, S. & Tabor, G. (2006) The use of computational fluid dynamics in reconstructing the hydrodynamic properties of graptolites. *GFF*, **128 (2)**, 189–194.

Rohlf, F.J. (1999) Shape statistics: procrustes superimpositions and tangent spaces. *Journal of Classification*, **16 (2)**, 197–223.

Rohlf, F.J. & Marcus, L.F. (1993) A revolution in morphometrics. *Trends in Ecology and Evolution*, **8 (4)**, 129–132.

Scott, R.S., Ungar, P.S., Bergstrom, T.S., et al. (2005) Dental microwear texture analysis shows within-species diet variability in fossil hominins. *Nature*, **436 (7051)**, 693–695.

Scott, R.S., Ungar, P.S., Bergstrom, T.S., et al. (2006) Dental microwear texture analysis: technical considerations. *Journal of Human Evolution*, **51 (4)**, 339–349.

Sellers, W.I. & Crompton, R.H. (2004) Using sensitivity analysis to validate the predictions of a biomechanical model of bite forces. *Annals of Anatomy*, **186 (1)**, 89–95.

Sellers, W.I., Hepworth-Bell, J., Falkingham, P.L., et al. (2012) Minimum convex hull mass estimations of complete mounted skeletons. *Biology Letters*, **8 (5)**, 842–845.

Shang, J.S. (2004) Three decades of accomplishments in computational fluid dynamics. *Progress in Aerospace Sciences*, **40 (3)**, 173–197.

Shiino, Y., Kuwazuru, O. & Yoshikawa, N. (2009) Computational fluid dynamics simulations on a Devonian spiriferid *Paraspirifer bownockeri* (Brachiopoda): generating mechanism of passive feeding flows. *Journal of Theoretical Biology*, **259 (1)**, 132–141.

Shiino, Y. & Kuwazuru, O. (2010) Functional adaptation of spiriferide brachiopod morphology. *Journal of Evolutionary Biology*, **23 (7)**, 1547–1557.

Shiino, Y., Kuwazuru, O., Suzuki, Y., et al. (2012) Swimming capability of the remopleuridid trilobite *Hypodicranotus striatus*: hydrodynamic functions of the exoskeleton and the long, forked hypostome. *Journal of Theoretical Biology*, **300**, 29–38.

Souter, T., Cornette, R., Pedraza, J., et al. (2010) Two applications of 3D semi-landmark morphometrics implying different template designs: the theropod pelvis and the shrew skull. *Comptes Rendus Palevol*, **9 (6–7)**, 411–422.

Teaford, M.F. & Walker, A. (1984) Quantitative differences in dental microwear between primate species with different diets and a comment on the presumed diet of *Sivapithecus. American Journal of Physical Anthropology*, **64 (2)**, 191–200.

Tseng, Z.J. (2009) Cranial function in a late Miocene *Dinocrocuta gigantea* (Mammalia: Carnivora) revealed by comparative finite element analysis. *Biological Journal of the Linnean Society*, **96 (1)**, 51–67.

Ungar, P.S., Brown, C.A., Bergstrom, T.S., et al. (2003) Quantification of dental microwear by tandem scanning confocal microscopy and scale-sensitive fractal analysis. *Scanning*, **25 (4)**, 185–193.

Ungar, P.S., Scott, R.S., Grine, F.E., et al. (2010) Molar microwear textures and the diets of *Australopithecus anamensis* and *Australopithecus afarensis. Philosophical Transactions of the Royal Society B*, **365 (1556)**, 3345–3354.

Walker, A., Hoeck, H.N. & Perez, L. (1978) Microwear of mammalian teeth as an indicator of diet. *Science*, **201 (4359)**, 908–910.

Witmer, L.M. (1995) The extant phylogenetic bracket and the importance of recon-structing soft tissues in fossils. In: Thomason, J.J. (ed), *Functional Morphology in Vertebrate Paleontology*, pp. 19–33. Cambridge University Press, Cambridge.

Zienkiewicz, O.C. (1971) *The Finite Element Method in Engineering Science*. McGraw-Hill, London.

Further Reading/Resources

Curtis, N. (2011) Craniofacial biomechanics: an overview of recent multibody modelling studies. *Journal of Anatomy*, **218 (1)**, 16–25.

MacLeod, N. (2002) Geometric morphometrics and geological shape-classification systems. *Earth-Science Reviews*, **59 (1–4)**, 27–47.

O'Higgins, P., Fitton, L.C., Phillips, R., et al. (2012) Virtual functional morphology: novel approaches to the study of craniofacial form and function. *Evolutionary Biology*, **39 (4)**, 521–535.

Pozrikidis, C. (2011) *Introduction to Theoretical and Computational Fluid Dynamics*. Oxford University Press, New York.

Rayfield, E.J. (2007) Finite element analysis and understanding the biomechanics and evolution of living and fossil organisms. *Annual Review of Earth and Planetary Science*, **35**, 541–576.

Richmond, B.G., Wright, B.W., Grosse, I., et al. (2005) Finite element analysis in functional morphology. *The Anatomical Record Part A*, **283A (2)**, 259–274.

Shabana, A.A. (2005) *Dynamics of Multibody Systems*. Cambridge University Press, New York.

Ungar, P.S. (2009) Tooth form and function: insights into adaptation through the analysis of dental microwear. In: Koppe, T., Meyer, G. & Alt, K.W. (eds), *Comparative Dental Morphology*, pp. 38–43. Karger, Basel.

Zelditch, M.L., Swiderski, D.L. & Sheets, H.D. (2004) *Geometric Morphometrics for Biologists: A Primer*. Elsevier Academic Press, New York.

7

Summary

Abstract: Virtual palaeontology techniques are now mainstream, but no overview of them all has yet been published; a summary is hence provided here. A hierarchical taxonomy of techniques is proposed, and semi-quantitative comparisons of data-capture techniques are presented. Recommendations for the selection of data-capture methods are also presented in the form of a flow chart, and recommendations for visualization methods are briefly outlined. Trends in the usage of techniques and likely future developments are discussed.

7.1 Introduction

Palaeontologists use virtual specimens for many purposes. The most important of these has always been the exploration and description of morphology, taking advantage of the abilities of the methods described in this book to elucidate internal details, extract specimens from matrix, magnify details and perform dissections. Intimately related to this is the dissemination of digital visualizations (static images, stereo-pairs, pre-rendered videos, and interactive reconstructions) as an accompaniment to scientific publications (Chapter 5). In addition, quantitative analyses of form and function are increasingly making use of this data (Chapter 6), and virtual fossils (and physical models produced by three-dimensional printing) have also proven effective tools for communicating results beyond the scientific community (Chapter 5). For all these reasons, the use of virtual fossils as proxies for real specimens is no longer a rare or specialized approach in palaeontology; in the course of little more than 10 years, it has become mainstream.

Despite the increasing acceptance and uptake of these methods, an easy-to-digest synopsis of the underlying methodologies and their capabilities, weaknesses and applicabilities has hitherto been lacking. To supplement the detailed treatments of provided in Chapters 2–5, this chapter offers a

Techniques for Virtual Palaeontology, First Edition. Mark D. Sutton,
Imran A. Rahman and Russell J. Garwood.
© 2014 John Wiley & Sons, Ltd. Published 2014 by John Wiley & Sons, Ltd.

summary of these approaches (Section 7.2), recommendations for method selection (Section 7.3), and analysis of trends in their use, together with speculation on the near future (Section 7.4).

7.2 Summary of Data-Capture Methodologies

Chapters 2–4 describe a (perhaps bewildering) array of approaches to digitizing three-dimensional fossil form. These are best understood with a simple taxonomy (Figure 7.1). The most fundamental division is between surface-based and tomographic techniques, the former capturing only the exposed surface, and the latter additionally providing data on the internal structures of suitable specimens. Tomographic techniques, in turn, can be divided into destructive and non-destructive approaches, the former including traditional approaches such as serial grinding, and the latter encompassing scanning technologies such as CT.

The vast majority of palaeontological specimens that preserve a degree of three-dimensionality will be amenable to at least one of these methods. Table 7.1 summarizes the most important properties and limitations of the techniques for data capture discussed in this book and provides references to the sections where each method is discussed in more detail. Figure 7.2 provides a visual guide to the scales at which these methods are capable of operating.

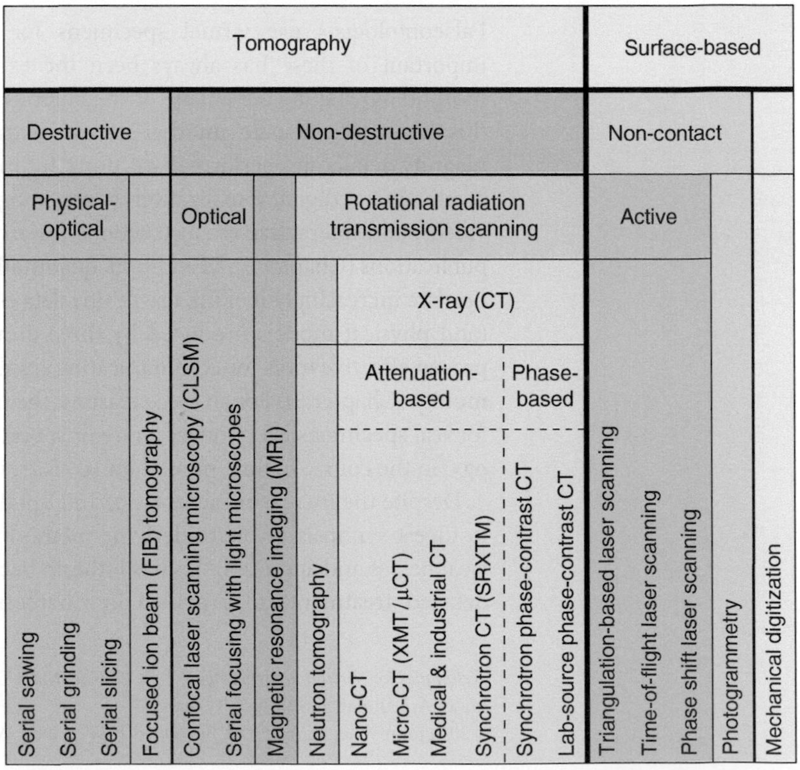

Figure 7.1 A simple taxonomy of data-capture methodologies.

Table 7.1 Semi-quantitative comparison of different techniques.

Technique	Surface or tomographic (T/S)[*]	Destructive? (Y/N/Semi)[*]	Images what?	Registration needed? (Y/N/S)[†]	Acquisition speed	Size of feature of interest	Finest detail resolvable[‡]	Equipment availability	Equipment cost, or (cost per specimen)[§]	Portable? (Y/N/ Semi)[¶]	Difficulty (High/ Med/ Low)[**]	Section
Physical-optical (grinding)	T	Y	Optical data from ground surfaces	Y	Days to weeks	1–100 mm	10 μm	Common	£100s to £10,000s	S	H	2.2.1.1
Physical-optical (sawing)	T	S	Optical data from saw-cut surfaces	Y	Hours to days	50 mm and up	250 μm	Uncommon	£1000s to £10,000s	S	H	2.2.1.2
Physical-optical (slicing)	T	S	Optical data from cut slices	Y	Days to weeks	1–30 mm	15 μm	Common	£100s to £1000s	S	H	2.2.1.3
Focused ion beam tomography (FIB)	T	Y	SEM images of milled surfaces (distorted?)	Y	Hours to days	1–500 μm	10 nm	Uncommon	(£100s)	N	H	2.3
X-ray micro-tomography (XMT/μCT)	T	N	X-ray attenuation	N	Minutes to hours	500 μm to 250 mm	1 μm	Fairly common	£10,000s to £100,000s (£10s to £100s)	N	M	3.2.4
Medical & industrial CT	T	N	X-ray attenuation	N	Minutes	>200 mm	100 μm	Medical common	(£10s to £100s)	N	L	3.2.5
Nano-CT	T	N	X-ray attenuation	N	Hours to days	1–60 μm	50 nm	Uncommon	(£10s to £100s)	N	H	3.2.6
Synchrotron CT	T	N	X-ray attenuation	N	Minutes	100 μm to 30 mm	50 nm	Rare	N/A	N	M	3.2.7
Synchrotron phase-contrast CT	T	N	X-ray phase modification (refraction)	N	Minutes	100 μm to 60 mm	100 nm	Rare	N/A	N	M	3.2.10

(continued)

Table 7.1 (*contd*)

Technique	Surface or tomographic (T/S)	Destructive? (Y/N/Semi)*	Images what?	Registration needed? (Y/N/S)†	Acquisition speed	Size of feature of interest	Finest detail resolvable‡	Equipment availability	Equipment cost, or (cost per specimen)§	Portable? (Y/N/Semi)¶	Difficulty (High/Med/Low)**	Section
Lab-based phase-contrast CT	T	N	X-ray phase modification (refraction)	N	Minutes to hours	500 μm to 250 mm	800 nm	Rare	(£10s to £100s)	N	H	3.2.10
Neutron tomography (NT)	T	N	Neutron attenuation	N	Minutes to hours	2–300 mm	30 μm	Rare	N/A	N	M	3.3
Magnetic resonance imaging (MRI)	T	N	Distribution of light elements (especially H)	N	Minutes to days	<1 m	10 μm	Non-medical is uncommon	N/A	N	M	3.4
Confocal laser-scanning microscopy (CLSM)	T	N	Reflected or fluorescing light through translucent materials	N	Minutes to hours	10–250 μm	300 nm	Uncommon	(£10s to £100s)	N	M	3.5.2.1
Serial focus with light microscope	T	N	Reflected light through translucent material	Y	Hours	100 μm to 10 mm	100 nm	Common	£100s to £1000s	S	M	3.5.4
Triangulation-based laser scanning	S	N	Topography and colour of surface	S	Minutes to hours	1 mm to 1 m	50 μm	Common	£100s to £10,000s	Y/S	L	4.2.2.1
Time-of-flight laser scanning	S	N	Topography and colour of surface	S	Minutes	>10 mm	1 mm	Fairly common	£10,000s	S	L	4.2.2.2

Phase-shift laser scanning	S	N	Topography and colour of surface	Seconds to minutes	>10 mm	1 mm	Uncommon	£10,000s	S	L	4.2.2.3
Photo-grammetry	S	N	Topography and colour of surface	Minutes to hours	Any	N/A	Very common	£10s to £1000s	Y	L	4.3
Mechanical digitization	S	N	Topography of surface	Minutes to hours	>50 mm	1 mm	Uncommon	£100s to £1000s	S	M	4.4

All data should be taken as indicative estimates.

*Semi-destructive approaches are those where some material is destroyed, but the majority is retained.

†Registration of surface datasets (Section 5.3) is recorded as 'S'; this differs fundamentally from tomographic registration (Section 5.2.2), which is recorded as 'Y'.

‡For tomographic approaches that are not isotropic, the highest (worst) figure (typically tomogram spacing) is given.

§For some methods, equipment cost is not important as normal usage involves obtaining time on third-party equipment; here an estimated cost per specimen is given in parentheses. Time on infrastructural equipment (e.g. synchrotrons) is normally not charged; instead potential users must apply for beam time. Such cases are listed as N/A in the preceding, as are experimental or rare systems whose potential charging policy we cannot comment on.

¶Portable equipment can be easily carried by a single person, potentially into the field. Semi-portable equipment is 'luggable' from one laboratory to another.

**Our qualitative estimate of how complex and challenging the entire procedure will be to a relative novice; complications with visualization are included in the estimate, as are likely degrees of technical support available.

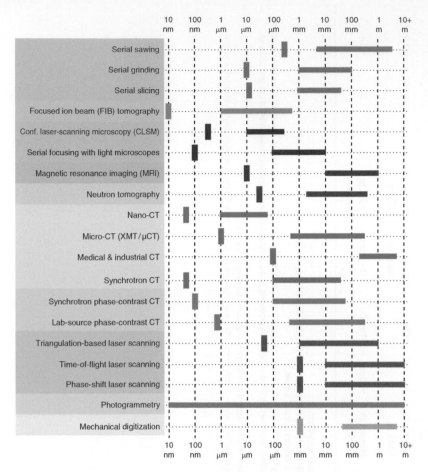

Figure 7.2 Visual comparison of scales at which different methods of Figure 7.1 are capable of working. Thick lines represent approximate range of feature-of-interest sizes. Vertical bars represent approximate minimum resolvable feature size. Note that photogrammetry is essentially scale-agnostic and the minimum size given (for SEM photogrammetry) is notional.

7.3 Recommendations for Method Selection

Selecting the optimal method for three-dimensional data capture is not straightforward; it will depend on both the properties of the specimen(s) to be studied, and an understanding of the methods in question. Ideally, method selection would follow detailed study of this book, coupled with further reading of relevant literature. We recognize, however, that this might sometimes be unrealistic, and that there is some value in rule-of-thumb recommendations for methodologies. In this spirit, Figure 7.3 provides recommendations for data-capture methodologies in the form of a flow chart. It is intended only as a guide, but captures the prioritization of methods that we would adopt for specimens of different sizes, X-ray amenability and translucency.

Visualization method selection is not covered in Figure 7.3, but our broad recommendations are relatively straightforward. Non-contact surface-based

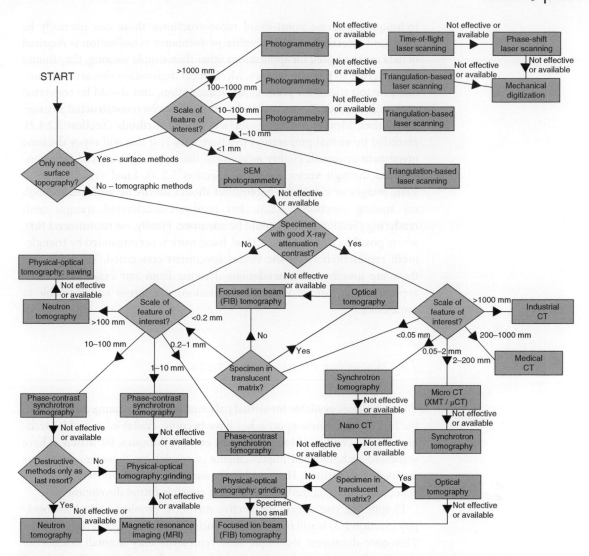

Figure 7.3 Flow chart showing our recommendations for technique selection for three-dimensional data capture. We stress that these are no more than recommendations, and that the use of this flow chart is not a substitute for a full understanding of the capabilities of the methods detailed in this book. Green triangular boxes represent decisions, and have multiple exits; orange/pink rectangular boxes represent recommendations that a particular method should be considered. Most methods have a complex set of requirements, strengths and weaknesses, detailed in the appropriate section of this book. Not all will be applicable to all specimens – for example, neutron tomography is particularly applicable to specimens preserving organic material, and performs less well where organics are absent. Most method-recommendation boxes have an exit marked 'not effective or available', which should be followed if the method is tried and fails for any reason, or is simply not practical (e.g. synchrotron beam time may not be available). Where a methods box has no 'not effective or available' exit, but the method still fails, then that specimen may not be amenable to any form of data capture.

techniques generate point-cloud reconstructions; these can normally be visualized directly, but where higher performance visualization is required or data are to be used for applications other than simple viewing, they should be converted to triangle meshes. Mechanical digitization datasets will normally be too sparse for point-cloud visualization, and should be converted to triangle meshes. Tomographic data should also be reconstructed as triangle meshes, ideally through isosurface volume methods (Section 5.2.4.2), preceded by virtual preparation (Section 5.2.4.3) if this will repay the time investment required. Highly anisotropic datasets might instead be reconstructed through vector surfacing (Section 5.2.3). Final visualization of static images or videos for publication should ideally be achieved through ray tracing (Section 5.4.2.3), but hardware-accelerated triangle-mesh rendering (Section 5.4.2.2) could be adequate. Finally, we recommend that, where possible, all published virtual-fossil work is accompanied by triangle-mesh representations of the virtual specimens concerned. We stress that these are merely recommendations deriving from our experience; other approaches (e.g. direct volume visualization) may prove equally viable in some cases.

7.4 Developments and Trends

The techniques available for virtual palaeontology have changed radically in the last 20 years; new approaches have become available, and older techniques have continued to evolve. Most or all techniques, for instance, have witnessed technological improvements in resolution; these improvements are likely to continue. Here we briefly analyze other recent trends in data collection and visualization, and discuss possible future developments.

Despite the rise of non-destructive methods, physical-optical tomography (Section 2.3) is still in use, almost entirely in the form of serial grinding. This once-dominant technique now represents only a small fraction of virtual palaeontological work, but this more reflects the growth of virtual palaeontology as a sub-discipline than the atrophy of this traditional method. In part, this continuity represents inertia, but physical-optical tomography does retain advantages over non-destructive approaches for certain types of material (i.e. fossils that lack internal X-ray contrast), and it is likely to continue, at the very least as a method of last resort, for the foreseeable future. Focused Ion Beam (FIB) tomography (Section 2.3), the other destructive method discussed herein, is – in contrast – very much in the early stages of uptake in palaeontology. Whilst the technique will likely remain time-consuming and challenging, it provides the highest resolution of any method we discuss; it hence has great potential for the study of very small microfossils, or very fine features (e.g. histology, micro-structure) of larger, exceptionally preserved samples. That this high resolution can theoretically be augmented by three-dimensional elemental mapping greatly enhances the potential of the approach; we anticipate a substantial growth in its use in the future.

X-ray computed tomography (CT) is currently the mainstay of virtual palaeontology. It is applicable to the vast majority of preservational modes (especially when augmented by phase-contrast techniques), has variants that cover a very wide range of scales, and is – for the most part – cheap and accessible. It also produces relatively clean and easily handled datasets, although these can be unwieldy in size. Of the variants of CT discussed in Section 3.2, industrial, medical and micro-CT are now mature and widely available technologies; their capabilities are increasingly well-understood by the palaeontological community. Nonetheless, they are still not as widely used as they might be, particularly outside the vertebrate palaeontology community; as micro-CT systems continue to fall in price we anticipate a further broadening of their use. Nano-CT is a relatively new technology, and its palaeontological usage has been limited to date. It may become significant in the near future, particularly for the study of microfossils. Synchrotron-based CT has become the most high-profile variant of the technology, and can often (although not always) provide the cleanest and highest-resolution datasets available, particularly for small specimens. The limited capacity of palaeontologically-suitable synchrotron beamlines may restrict the growth of these techniques, although decreasing scan-times could partially mitigate this effect.

Developments in CT go beyond increases in resolution and availability. The increasing use of phase-contrast methods (Section 3.2.10), most notably at synchrotrons but also now using lab-sources, is an important development that has greatly increased the resolving power of CT for difficult (low attenuation-contrast) specimens. Additionally, the development of new methods that are capable of mapping elemental or mineralogical composition in three dimensions (Section 3.2.12) are of great potential significance to palaeontologists; colour CT is perhaps the most exciting of these. Such a three-dimensional map could provide not only improved anatomical resolution, but also revolutionize studies of taphonomy. Non-destructive compositional tomography methods are still experimental and/or prohibitively time-consuming; should they become more widely available and practical to apply, we predict a very significant palaeontological uptake.

Neutron tomography (NT; Section 3.3) and magnetic resonance imaging (MRI; Section 3.4) have hitherto been used only occasionally in palaeontology, as they are generally less accessible and generate lower-resolution datasets than CT. While they may have some utility for specimens not amenable to CT, especially in the case of organic-rich fossils which often respond well to NT, we do not anticipate any rapid growth in their palaeontological uptake in the foreseeable future. Optical tomography (Section 3.5) has also seen only occasional use, but here we see more potential for expansion. This approach is limited to small specimens in translucent materials (e.g. fossils preserved in non-opaque cherts), but where these requirements are met it is an attractive proposition, capable of recovering fine-detail resolution from otherwise difficult material. We expect to see it develop into the method of choice for the (admittedly limited) set of specimens to which it can be applied.

Surface-based techniques have seen a rapid increase in palaeontological usage in recent years, the vast majority of such studies using laser scanning (Section 4.2). This approach is relatively cheap and simple to use, is capable of capturing colour data, and provides adequate resolution of detail for most specimens. Photogrammetric techniques (Section 4.3), however, are currently undergoing very rapid development; while palaeontological studies to date have been few, the attractions of this approach include its scale-agnostic nature, high portability and very low equipment costs. Photogrammetric reconstructions can often rival or even exceed the accuracy of laser scanning. As many palaeontological applications do not require the imaging of the internal structures of specimens, we anticipate a surge in interest in photogrammetry, which may soon become the most common way to digitize fossils. Mechanical digitization (Section 4.4) has seen few applications to date, and we do not anticipate any future rise in its popularity as means of producing virtual specimens; modern non-contact methods are generally more attractive.

The visualization of tomographic data has witnessed a steady decline in the use of older vector-surfacing approaches (Section 5.2.3), which have been replaced with volume-based techniques, principally the calculation of isosurfaces (Section 5.2.4.2). We do not anticipate any reversal of this trend. Direct volume rendering (Section 5.4.3), while theoretically a viable alternative to isosurfaces, has never been widely used in palaeontology; we see no reason to expect this to change. Increases in hardware capability might be expected to improve availability of interactive rendering, but in the past these gains have typically been offset by increases in the resolution and hence size of reconstructions; we hence do not predict any substantial change in the (already satisfactory) accessibility of rendering hardware. Software for reconstructions is likely to continue to be refined, although this process will likely be slow, and there is no reason to expect any reduction in price for commercial packages. Data interchange in virtual palaeontology would be greatly facilitated by developments of standardized file formats and the availability of online repositories for digital fossils (Section 5.5.2). The speed and direction of any such developments is difficult to predict, but we are hopeful that moves to standardize file formats at least will come to fruition in the near future. One further obvious prediction relates to three-dimensional printing (Section 5.4.2.4); the technologies involved are becoming increasingly accessible, and we expect a rapid increase in the number of virtual reconstructions reproduced in this way. Typically, this would be as an accessory to, rather than a replacement for, on-screen visualization.

As digital reconstructions of fossils become increasingly common, the number and frequency of applications beyond visualization will rise. The vertebrate palaeontological community has been most active in utilizing quantitative methods to analyze the form and function of virtual fossils – especially geometric morphometrics (Section 6.2) and finite-element analysis (Section 6.4.1). We anticipate a steady increase in the usage of such methods in vertebrate palaeontology and a more rapid rise in non-vertebrate palaeontology; here, computational fluid dynamics (Section 6.4.4) holds considerable promise (e.g. simulating fluid flow past reconstructions of marine invertebrates and air flow past virtual seeds). However, uptake of biomechanical modelling (Section 6.4) will also depend on the availability of experimental data for modern species, required to generate model parameters and to validate computer models.

7.5 Concluding Remarks

The techniques which we describe in this book are of immense significance to the science of palaeontology; we have been privileged to have witnessed this virtual palaeontological revolution from within, and on occasions to have even helped direct it. The present work represents a distillation of over 25 years combined experience with techniques for virtual palaeontology; we hope and intend that it will create a broader understanding and encourage the evermore fruitful use of these approaches. These methods are not entirely novel, representing instead the modernization, broadening and refinement of the long-standing palaeontological tradition of 'serial sectioning'. Nonetheless, they provide a vastly expanded toolkit for the study of three-dimensional fossils, in ways which would have been impractical – if not impossible – in the recent past. As such, they are able to help us extract fresh data from otherwise intransigent specimens, and hence obtain new insights. Mastery of any one of these techniques, or even familiarity with them, has hitherto been considered something of specialism within palaeontology. We contend that their importance is now such that all palaeobiologists require at the very least a passing knowledge of them, and that a more in-depth understanding should be one of the core skills that we impart to palaeobiologists of the future.

Glossary

Acceleration voltage: The potential difference between cathode and anode in a non-synchrotron X-ray source, which helps define the X-ray energy.

Acetate peels: See Peels.

Anaglyph stereo: A visualization approach where images from two closely-spaced viewing positions are coloured red and green/cyan and can hence be viewed stereographically (i.e. three-dimensionally) using red/green or red/cyan coloured glasses.

Analytical reconstruction methods: A group of computed tomography (q.v.) reconstruction algorithms which are computationally efficient, but prone to artefacts.

Artefacts: In any form of scanning, an artefact is a systematic discrepancy between the reconstructed tomogram (q.v.) and the sample's true attenuation (q.v.) coefficients.

Attenuation: The loss of intensity of incident radiation as it passes through a medium. This typically results from absorption and/or scattering.

Attenuation coefficient: A measure of the strength of attenuation (q.v.) of a material per unit length.

Automatic first-pass registration: Using automated registration (q.v.) prior to manual registration (q.v.) to perform tomographic registration (q.v.).

Automatic registration: Using an automated algorithm to perform tomographic registration (q.v.).

Beam hardening: An artefact (q.v.) found in computed tomography scans with a polychromatic source (q.v.). Beam hardening results from differential attenuation of X-rays at different energies.

Binning: See Downsampling.

Body-size estimation: A method that uses computer models (q.v.) to reconstruct the dimensions of an extinct organism.

Boundary conditions: In a computer model, these are the loads and constraints that are applied to the model.

Bremsstrahlung: A continuous curve between the minimum and maximum X-ray energies in an X-ray source. Punctuated by characteristic radiation (q.v.).

Brilliance: A measure of synchrotron beam quality based on the number of photons emitted per second, the beam's collimation, the source area, and the spectral distribution.

Calibration images: Also known as flat and dark fields, these are images used to correct projections (q.v.) prior to filtered backprojection (q.v.) in a computed tomography (q.v.) scan.

Techniques for Virtual Palaeontology, First Edition. Mark D. Sutton, Imran A. Rahman and Russell J. Garwood.

Cellulose acetate peels: See Peels.

Characteristic radiation: Energy peaks in the bremsstrahlung (q.v.) found in lab sources, which are unique to any given target metal (q.v.).

Colour CT: An experimental technique in which a specialized detector capable of resolving an X-ray spectrum for each pixel is used to collect projections (q.v.).

Compton scattering: The collision of an X-ray photon and an electron, when the energy of the former greatly exceeds the latter. This results in partial photon energy loss (then scattering or deflection) and a free secondary electron.

Computational fluid dynamics (CFD): A method that uses computer models (q.v.) to simulate flow around an object.

Computed tomography (CT): A form of tomography (q.v.) where tomograms (q.v.) are recovered indirectly via computation from projections (q.v.) acquired with the aid of penetrating radiation, rather than direct imaging.

Computer model: A computer program used to approximate the behaviour of a real-world system of interest.

Confocal laser-scanning microscopy: A form of confocal microscopy (q.v.) that uses a laser beam to image the sample.

Confocal microscopy: A method for serial focusing (q.v.) that images focal planes (q.v.) by eliminating out-of-focus light.

Confocal Raman imagery: The combination of confocal microscopy (q.v.) and Raman spectroscopy (q.v.) to map the chemical structure of a sample in three dimensions.

Cropping: Reducing the size of an image or volume so that it only contains a region of interest (q.v.).

CT revolution: The rapid uptake of virtual palaeontology (q.v.) based on the increasing availability of X-ray computed tomography (q.v.) in the early 21st century.

Dark field: A calibration image (q.v.) collected with the beam off.

Decimation: The process of algorithmically reducing the number of triangles in a triangle mesh (q.v.) while minimizing damage to the geometry of the represented object. See also Quadric error metric algorithms.

Dental microwear: The microscopic wear patterns on teeth.

Dental microwear texture analysis: A technique for quantitatively analysing dental microwear (q.v.) through the measurement of three-dimensional surface texture variables.

Destructive tomography: Encompasses all forms of tomography (q.v.) in which all or part of the specimen is destroyed during the exposure of physical tomographic surfaces.

Direct point-cloud rendering: See Direct point-cloud visualization.

Direct point-cloud visualization: Techniques for the visualization of a three-dimensional point cloud as a two-dimensional image or sequence of images, without requiring the generation of a triangle mesh (q.v.).

Direct volume rendering: Visualizing a volume (q.v.) directly, that is without generating a triangle-mesh (q.v.).

Downsampling: Reduction in the resolution of a volume, normally by an integral factor.

Fiduciary markings: Markings in a tomographic dataset (q.v.), typically from physical-optical tomography (q.v.), which exist to aid the process of registration (q.v.).

Filament current: The current running through the filament of a non-synchrotron X-ray source, which helps define the X-ray energy.

Filtered back projection: A common algorithm for tomographic reconstruction (q.v.) in which projections are filtered and then superimposed at their acquisition angle over a square grid.

Finite-element analysis (FEA): A method that uses computer models (q.v.) to reconstruct stress, strain and deformation in a structure.

Flat field: A calibration image (q.v.) collected with the beam on but no specimen between source and detector.

Focal plane: The plane through a sample that is in focus.

Focused ion beam (FIB) tomography: A form of destructive tomography (q.v.) in which tomographic surfaces are physically exposed using a focused beam of ions and then imaged with the ion beam or a coupled-electron beam.

Fresnel Zone Plates: A means of focussing X-rays through a series of rings via diffraction.

Geometric morphometrics: A method that uses landmarks (q.v.) to quantitatively analyse form.

Hard X-rays: X-rays with wavelengths between 0.01 nm (124 keV) and 0.1 nm (12.4 keV).

Hardware-accelerated triangle-mesh rendering: See Triangle-mesh rendering.

Island removal: The process of algorithmically removing portions of a triangle mesh (q.v.) that are disconnected from the main mesh.

Isosurface: A mathematically defined surface calculated from a volume (q.v.), following points of a constant value. Isosurfaces are typically calculated using the marching cubes algorithm (q.v.).

Iterative reconstruction algorithms: A group of computed tomography (q.v.) reconstruction algorithms which are computationally inefficient, but less prone to artefacts (q.v.) than analytical reconstruction methods (q.v.).

K-alpha doublet: Characteristic radiation (q.v.) that represents electron transitions from a p-orbital of the L-shell to the vacated K-shell.

K-edge: A jump in the attenuation coefficient (q.v.) of an element when X-ray energy exceeds the binding energy of an atomic electron.

K-edge subtraction: A means of three-dimensional elemental mapping using scans taken just above and below a K-edge (q.v.) which are then subtracted.

Kerf: The material removed by a saw cut or the width of that material.

Labels: See Masks.

Laminography: A form of X-ray tomography (q.v.) for highly anisotropic (i.e. flat) specimens.

Landmarks: Co-ordinates of points representing anatomical features.

Laser scanning: A surface-based technique (q.v.) that uses a laser beam to acquire numerous point co-ordinates for an object or area, which define a three-dimensional point cloud (q.v.).

Magnetic resonance imaging (MRI): A form of non-destructive tomography (q.v.) in which tomograms (q.v.) are produced by using magnetic fields to map the distribution of atomic nuclei in a sample. See also Nuclear magnetic resonance (NMR).

Manual registration: Performing tomographic registration (q.v.) manually, that is judging and adjusting correct registration for each tomogram (q.v.) by eye, normally with the aid of fiduciary markings (q.v.).

Marching cubes: An algorithm for the calculation of an isosurface (q.v.), which generates a surface in the form of a triangle mesh (q.v.).

Masks: Regions of a volume (q.v.) flagged in software as belonging to a particular structure or region; masks are normally specified to enable selective deletion or differential rendering (e.g. colouring) of items in a volume. Also called labels or segments in some software.

Material properties: In a computer model, these are the physical properties (e.g. density and elasticity) that are assigned to the different materials (e.g. bone) in the model.

Mechanical digitization: A surface-based technique (q.v.) that uses the position of a three-dimensional digitization stylus held against a specimen to digitize its form.

Monochromatic: Of an X-ray or other electromagnetic source – comprising a single wavelength, for example a synchrotron (q.v.) beam passed through a monochromator. See also Polychromatic.

Multibody dynamics analysis: A method that uses computer models (q.v.) to simulate the movements of interconnected objects.

Nanotomography (nano-CT): A form of X-ray computed tomography (q.v.) in which sub-micrometre voxel-sizes are attained.

Neutron tomography: A form of non-destructive tomography (q.v.) in which projections (q.v.) are produced by exposing a sample to a beam of neutrons and recording the resulting neutron attenuation, with subsequent computational analysis to create tomograms (q.v.).

Non-destructive tomography: Encompasses all forms of tomography (q.v.) which do not require physical removal of portions of the specimen. See also Destructive tomography.

Nuclear magnetic resonance (NMR): A physical phenomenon in which nuclei in a magnetic field absorb and re-emit electromagnetic radiation. The basis of Magnetic resonance imaging (q.v.).

Optical tomography: A form of non-destructive tomography (q.v.) in which tomograms (q.v.) are produced by shining light through a sample.

Peels: A method for providing a permanent record of a surface (e.g. a tomographic surface) by chemically impregnating a cellulose peel with material from the surface.

Phase-contrast tomography: A form of X-ray computed tomography (q.v.) which maps the phase shift caused by a sample (and hence its refractive index) rather than a beam's attenuation.

Phase-shift laser scanning: A form of laser scanning (q.v.) that measures the distance between the scanner and the object by comparing the change in phase between the emitted and reflected laser light.

Photoelectric effect: A form of attenuation (q.v.) where an X-ray photon's energy slightly exceeds the binding energy of an atomic electron liberating it as a photoelectron (q.v.).

Photoelectron: An electron liberated by the photoelectric effect (q.v.).

Photogrammetry: A surface-based technique (q.v.) that uses multiple static images of a specimen, captured from different relative positions, to generate a three-dimensional virtual model.

Physical modelling: The production of a physical reproduction of a specimen from a virtual palaeontology dataset, or (historically) from analogue equivalents.

Physical-optical tomography: A form of destructive tomography (q.v.) in which tomograms (q.v.) are produced by physical exposure of surfaces and optical imaging or tracing.

Point cloud: A series of points in three-dimensional space used to define a surface; point clouds in virtual palaeontology (q.v.) are normally generated by surface-based techniques (q.v.), and each point may have a colour associated with it.

Polychromatic: Of an X-ray or other electromagnetic source – comprising multiple wavelengths, for example a lab-based X-ray source. See also Monochromatic.

Polygon mesh: A series of adjacent triangles or higher-order polygons defined by the three-dimensional co-ordinates of their vertices; triangle meshes (q.v.) are the form of polygon mesh normally used in virtual palaeontology (q.v.).

Projections: Two-dimensional images of a three-dimensional object in which incident radiation is differentially absorbed on the path through the sample. Traditional two-dimensional X-ray radiographs are examples of projections. Projections are used to create tomograms (q.v.) in computed tomography (q.v.).

Quadric error metric algorithms: Decimation (q.v.) algorithms that are capable of generating high-fidelity mesh simplification, but are often computationally expensive.

Raman spectroscopy: A technique for analysing the chemical composition of a sample.

Ray tracing: A computationally expensive but optically realistic means of generating a two-dimensional image from a three-dimensional geometry such as a triangle mesh (q.v.).

Region of interest: The portion of a volume that contains the specimen or portion of the specimen to be reconstructed.

Registered tomographic dataset: A tomographic dataset (q.v.) for which tomographic registration (q.v.) has been accomplished. Many scanning techniques (e.g. X-ray computed tomography, q.v.) produce datasets that are pre-registered, and do not require a discrete registration step.

Registration: The process of aligning data. The term is broad, and in virtual palaeontology (q.v.) can mean either (a) the aligning of two or more surface-based datasets or image stacks into a composite dataset, or (b) tomographic registration (q.v.).

Ring artefacts: An artefact (q.v.) which manifests as rings in a tomogram (q.v.), and results from variable sensitivities in detector elements.

Segments: See Masks.

Sensitivity analysis: An approach in computer modelling where input parameters are modified to evaluate their influence on the results and, hence, the uncertainty of the model.

Serial focusing: A form of optical tomography (q.v.) in which tomographic surfaces are captured by focusing a microscope at successive depths within a sample.

Serial grinding: A form of physical-optical tomography (q.v.) in which tomographic surfaces are physically exposed by grinding or lapping.

Serial sawing: A form of physical-optical tomography (q.v.) in which tomographic surfaces are physically exposed by saw cuts.

Serial slicing: A form of physical-optical tomography (q.v.) in which physical exposure of tomographic surfaces is accomplished by slicing with a blade.

Soft X-rays: X-rays with wavelengths between 0.1 nm (12.4 keV) and 10 nm (124 eV).

Spiral CT: The most common form of medical X-ray computed tomography (CT, q.v.) scanning in which a source and detector rotate around a sample moving in the z-direction.

Surface-based methods: Approaches to the gathering of data for virtual palaeontology (q.v.) that acquire topographic data on exposed surfaces of a specimen. All digitization techniques covered in this book are either surface based or use tomography (q.v.).

Synchrotron: A particle accelerator that uses charged particles (usually electrons) circulating in a storage ring to produce intense X-rays.

Synchrotron radiation X-ray tomographic microscopy (SRXTM): A high resolution form of synchrotron tomography (q.v.).

Synchrotron tomography: A form of X-ray computed tomography (q.v.) that employs a synchrotron (q.v.) as its X-ray source.

Target metal: The metal used to create X-rays by bombardment with electrons, in a laboratory X-ray source.

Three-dimensional printing: A type of physical modelling where a three-dimensional digital geometry, normally in the form of a triangle mesh (q.v.), is printed as a physical object.

Time-of-flight laser scanning: A form of laser scanning (q.v.) that measures the distance between the scanner and the object by calculating the time taken for the laser beam to return to the scanner.

Tomogram: An individual slice produced via tomography (q.v.).

Tomograph: A device used to perform tomography (q.v.).

Tomographic dataset: The dataset produced by tomography (q.v.) of a specimen.

Tomographic reconstruction: The creation of tomograms (q.v.) from projection (q.v.) data through a variety of algorithms such as filtered back projection (q.v.).

Tomographic registration: The process of aligning tomograms (q.v.) with respect to each other. Formally, this involves transforming images such that the vector offset in real space between any point (x,y) in tomograms n and n + 1 is perpendicular to the tomographic plane. See also Registration.

Tomography: The study of three-dimensional structures through a series of two-dimensional parallel slices through a specimen. All digitization techniques covered in this book either use tomography, or are surfaced based (q.v.).

Triangle mesh: A series of adjacent triangles defined by the three-dimensional co-ordinates of their vertices, used to provide a numerical representation of a surface. See also Polygon mesh.

Triangle-mesh rendering: The process of visualizing a three-dimensional triangle mesh (q.v.) as a two-dimensional image, normally using dedicated graphics hardware.

Triangulation-based laser scanning: A form of laser scanning (q.v.) that uses triangulation to determine point co-ordinates for a sample.

Validation: An approach in computer modelling where experimental data are compared with the model results to determine the accuracy of the model.

VAXML: A proposed standard for the storage and dissemination of virtual fossils (q.v.) in triangle-mesh (q.v.) format.

Vector surfacing: A method or group of methods for visualizing tomographic data, in which mathematically defined curves (splines) generated for each tomogram are 'surfaced' by some algorithm to produce a three-dimensional digital representation of the specimen.

Virtual fossil: A fossil reconstructed as a three-dimensional digital model.

Virtual palaeontology: The study of three-dimensional fossils through digital visualizations.

Virtual preparation: Manual or semi-automatic 'tidying' or marking-up (e.g. masking, q.v.) of a virtual palaeontology (q.v.) dataset so as to improve its utility.

Volume: A three-dimensional dataset in which data is held as an array of voxels (q.v.) representing the values of some property. A volume is the three-dimensional equivalent of a raster (bitmapped) image.

Volume-based reconstruction: A visualization of a tomographic dataset (q.v.) in which the data is treated as a volume (q.v.). Volume-based reconstruction can either use an isosurface (q.v.) or direct volume rendering (q.v.).

Volume ray-casting: The most common form of direct volume rendering (q.v.).

Volume rendering: See Direct volume rendering.

Voxel: A volume element, representing a value at a point within a volume (q.v.). A voxel is the three-dimensional equivalent of a pixel.

XANES tomography: An experimental technique allowing three-dimensional mapping of the chemical speciation of an element.

X-ray computed tomography: A form of non-destructive tomography (q.v.) in which projections (q.v.) are produced by exposing a sample to an X-ray beam and recording the resulting X-ray attenuation, with subsequent computational analysis to create tomograms (q.v.).

X-ray microtomography: A form of high-resolution X-ray computed tomography (q.v.) in which voxel-sizes smaller than ~50 μm are attained. See also Nanotomography (nano-CT).

Index

Techniques for Virtual Palaeontology, First Edition. Mark D. Sutton,
Imran A. Rahman and Russell J. Garwood.
© 2014 John Wiley & Sons, Ltd. Published 2014 by John Wiley & Sons, Ltd.

Printed and bound by CPI Group (UK) Ltd, Croydon, CR0 4YY

27/10/2024

14580355-0005